MANUAL OF PRACTICAL MANAGEMENT

For Third World rural development associations

VOLUME II
Financial Management

FERNAND VINCENT

INTERMEDIATE TECHNOLOGY PUBLICATIONS
on behalf of IRED • 1997

Intermediate Technology Publications
103–105 Southampton Row, London WC1B 4HH, UK

© IRED 1989

A CIP catalogue record for this book is available from the British Library

ISBN 1 85339 405 X

Printed in the UK by SRP Exeter

In order to obtain local development, development strength must be local...

It is not right that there are always people who ask and those who give... The present donor-beneficiary system is not endless.

To become independent, we need strong means of financing, organisation and rigorous management.

Mamadou Cissokho
A village leader of Senegal

INTRODUCTION

The manual of PRACTICAL MANAGEMENT is addressed primarily to grassroots managers and people responsible for conducting, managing and inspiring non-governmental development associations in the Third World. It is also useful for training administrators of small groups or associations and those who manage development projects.

Written in a simple and direct style, the handbook contains numerous examples and annexes, the result of the African/Asian experience.

To facilitate easier understanding, a glossary of unfamiliar technical terms is appended.

The manual seeks to <u>clarify</u> management <u>problems</u> identified by a key word with references in the index.

<u>Volume I</u> of this manual, deals with the organisation, administration and communication of an association. Some of the chapters from Volume I are:

- Animation and participation
- Environmental study
- Planning, programming and projects
- The organising and follow-up of activities
- Education, information, documentation
- Communication techniques
- The organisation and administration of an association
- External relations
- Control and evaluation

Activities organised by IRED during the past 4 years in Africa and Asia had a strong influence on the present edition of this book. The first version of this manual has been published in French by the Sahelian associations.

This book has been translated and adapted from French into English by Ms. Shyamalee Tudawe of the IRED Asian Regional Office in Colombo, Sri Lanka with the collaboration of Sunimal Fernando, Director and Laurence Lançon, IRED Geneva. Its publication has been funded by SOS Faim - Brussels with co-financing from the EEC (European Economic Community).

Do not hesitate to inform us of your suggestions to enable us to improve future editions which will thus permit us to deal better with the needs of the groups and rural associations in the Third World.

Fernand VINCENT
Secretary General of
the International
Network of IRED

CONTENTS

INTRODUCTION

Part I

FINANCING AN ASSOCIATION
AND ITS ACTIVITIES

Part II

SAVINGS AND CREDIT SYSTEMS

Part III

SIMPLE ACCOUNTS, BUDGETS AND CASH-FLOWS

Part IV

MANAGEMENT OF SMALL PROJECTS

Part V

ACCOUNTING SYSTEMS

Part VI

PROFIT AND LOSS ACCOUNTS

Part VII

FINANCIAL CONTROLS AND JUSTIFICATION
OF EXPENSES RELATED TO GRANTS

INDEX and GLOSSARY
See end of book

* * * * *
* * *
*

PART ONE

FINANCING AN ASSOCIATION AND ITS ACTIVITIES

A. The non-financial resources at hand

B. Internal financial resources

C. Possibilities of local financing

D. International fund-raising

E. How to create flexible funds or revolving funds

F. Planning for the end of aid

A. THE NON-FINANCIAL RESOURCES AT HAND

Before carrying out a project, first find out what resources are available or not and then mobilise the available resources.

There are three kinds of resources:

- human resources
- material and technical resources
- financial resources

Although it is often assumed that the solution to most problems is financial, it is not always so. The answer could well be found by mobilising other non-financial resources which have to be identified and evaluated.

Human Resources

Several resource persons can be found within the Association or in the village. There are leaders, innovators, and experts in a particular field: craftsmen (artisans), livestock-breeders or fishermen etc. All of them can pass on their knowledge to others if they are called upon to do so.

Human resources found within a village comprise technical, administrative and political persons: government officers, teachers, agriculturists or politicians ready to support the group's activities.

Refer to Annexe I.1: List of those who can support the Association

A very important human resource is the manpower of all the members of the Association and the population of the village: LABOUR.

All the above-mentioned human resources should be mobilised for the benefit of the development projects of the Association.

Material and Technical Resources

There are innovative ideas, know-how and technology in the minds of innovative villagers: blacksmiths, craftsmen, weavers, and persons involved in animal husbandry as well as groups of women and youth.

During the study of the environment, the Committee of the Association should make a list of equipment, machinery, tools, land, buildings, vehicles, dams, water tanks, wells, granaries, health clinics and other available material resources useful for the activities of the Association.

The role of the Committee will be to evaluate all the available resources of the village and its neighbourhood.

B. INTERNAL FINANCIAL RESOURCES

The success of the Association depends on the efforts its members are capable of making, using their own means.

The Association should group all internal resources, financial or otherwise, required to launch projects. Before approaching the international funding organisations for aid, the Association should fund from its own resources its first small project. This is evidence that the group has made its own efforts before requesting aid from outside agencies.

There are a number of potential internal financial resources. For example, the membership fees. At a General Assembly, this must be discussed and the amounts decided on. Some groups fix the membership fee at a relatively high level to be able to judge a new member's level of commitment. Thus, the Association ensures having a committed group of people as its members.

The Committee should design an individual membership card. The Secretary should maintain an up-to-date list of members' subscriptions.

The members can participate in the financing of the Association by making specific material contributions. They could either donate or loan livestock, premises, land, etc. to the Association; e.g. in order to make up the initial stocks of a rice bank, each member's family can be requested to bring a bag of rice. Then it will not be necessary to purchase stocks from outside. The silos, too, can be built with material collected from the village and this will result in very little money being spent. Thus, the project would be funded by the Association.

Another way of financing the activities of the Association is to sell collective field produce. For the efficient functioning of the group, each person will work his own land while collective work will be organised on land given over to the Association by the members.

The sale of these products will generate financial resources, a part of which could be invested either in new productive projects, or in developing infrastructure facilities useful to the community or for setting up reserves which will enable, when necessary, the development of other new projects.

Such services provided by the Association will enable the members to generate a production surplus and to start saving. Savings is the motive force of economic development, as it provides capital for investment.

C. POSSIBILITIES OF LOCAL FINANCING

The villagers and their local Associations generally do not make sufficient use of local financing possibilities for their own projects.

The best way to finance local development is to use local savings or available community cash.

Another solution is to mobilise the resources of villagers living outside in the city who enjoy a relatively higher level of income. They could subscribe directly to the local association of the village or form a separate association whose activities could be supported out of such urban cash-flows.

There are donations in kind which the wealthy, the prominent people and the elders could contribute to the group. These donations could be made in the form of grain, tools, buildings or land.

Also, there are many rich persons in the country who could be approached to make financial donations to the group.

In Asian countries, there are small organisations - both religious as well as secular - which help launch small-scale local projects. The response of such local donors to applications for support depends on the integrity of the group and the good reputation it has earned.

Once an Association is officially recognised, it could obtain some government support which could at first be free (supply of fertilizer, plants, trees, seeds). If the group is well organised, it could obtain financial aid from the government or have access to credit from local banks.

D. INTERNATIONAL FUND-RAISING

When all internal and local fund-raising resources have been exhausted and/or the finances are not sufficient for the funding of a programme or of a project, there is always the possibility of approaching an international funding organisation.

There are several governmental and non-governmental donor agencies represented locally through their representatives/own offices or their embassies.

Aid is linked with certain requirements and conditions. Before getting tied up through a verbal or written contract with a donor agency, the association should be familiar with these conditions. At a certain stage, dependence on aid should cease as otherwise the local group will be "eternally assisted".

Once an association has been formed, for it to be able to develop its initial infrastructure, aid is often necessary. Before building up its organisational and infrastructural base - for which aid is often needed - an association cannot expect to move into a higher level of developmental activity. However once it starts implementing medium and long-term projects, it should ensure that the running costs of such projects are self-financed and not financed with grants/aid.

1. Getting to know the donor

The documents requesting aid must be completed according to the norme and procedures of the donor - it is he who has the money and he who decides. It is thus, important to know:

- his correct name and address
- his ideology: why would he help
- how and through whom is he present in the country
- the kind of projects to which he gives priority
- what are his criteria for selection of projects
- the maximum amount that he can fund
- who should be contacted in this organisation
- the procedures to be followed
- the period of time needed to study and make decisions
- the period of time taken for disbursing funds after a positive decision is made
- the requirements for the monitoring of funds disbursed.

Refer to Annexe I.2: Criteria for selection of projects

Processing of a financial request

a) The document will be first submitted to a technical sub-committee which will in turn assign one of its members for a detailed analysis of the project. He will evaluate the request and report to the sub-committee.

b) Next, this information will be discussed by a committee which will make a definite decision whether to grant aid or not.

c) After that, if the particular donor works with funds given by his government, he will be obliged to obtain the approval of his own government.

d) It is only when all these barriers have been passed that the donor will be in a position to communicate his final decision by letter.

e) If it is a positive decision, an estimated cash-flow will have to be submitted to schedule cash remittances according to the requirements of the project. The cash-flow estimate should be included in the original aid request in order to avoid delays at this stage.

f) On receiving this cash-flow plan, the donor can transfer the first instalment (in 2 to 6 weeks).
Each subsequent disbursement will be made only after the donor receives accounts for funds disbursed up to that time. The final instalment will be remitted only after all the expenses of the project are accounted for.

2. The donor's point of view

For his part, the donor will want to evaluate WHETHER HE CAN PLACE HIS CONFIDENCE IN THE LOCAL ASSOCIATION.
Therefore, he should know:

- the name of the Association
- its history and experience
- about any other organisation which has already funded the Association (a donor does not like to be the first and only one to finance an association)
- if the project submitted for funding is worthwhile - whether the idea is good - does it fulfil a priority need - is the technical support satisfactory?
- whether a realistic plan has been prepared to carry out the work undertaken
- the importance of local participation, human investment, the financial support to be contributed by the people and by the organisation
- whether the proposed activities are feasible on economic, financial and social grounds and if the funds are granted what results can be expected
- what is the exact amount requested and what is the duration for which financial support is needed ?

3. The contents of a document requesting aid

a) <u>For an amount less than $1,000.-</u>

- the name of your project
- its technical description
- the group's efforts
- the results hoped for
- budgetary cost estimate for the requested funds

The funding of such a project could be obtained locally through an embassy, an NGO representation or a local donor (document 3-4 pages).

b) <u>For an amount between $1,000.- and $10,000.-</u>

Forward a document of 6-7 pages with a little more detail than mentioned above.

If funding is requested for equipment (a pump, a milling machine, etc.) attach a proforma invoice. Such funds should be available within the country.

c) <u>For an amount over $10,000.-</u>

If the request exceeds $10,000.-, the procedure is longer and more complicated:

- produce a detailed document requesting aid
- contact the representative of the donor
- have the donor's representative inspect the dossier (also possibly visit the site)
- obtain further information to answer his questions
- the donor's representative will examine the dossier and despatch it to the head office
- a technical team will inspect the dossier
- there may be a request for further information
- dossier to be completed and information to be sent again
- decision of the technical team and of the advisory board and, if needed, the approval of the government of the donor's country.
- the local representative of the group will be informed of the donor's decision

Take into account that such a procedure takes as long as 6 to 12 months before a final decision is made, and 7 to 15 months before the funds are disbursed. Hence, plan financial requests well in advance.

An application for financial aid should generally contain:

- a description of the Association and its past history
- name of the project
- objectives
- the anticipated results
- the rationale for that project
- organisation
- resources
- operational plan
- costs (budget)
- profitability
- total financing
- funds requested from the donor

Annexes I.3, I.4 & I.5 give concrete examples of documents requesting aid.

4. Categories of donors

There are several types of international donor agencies:

- non-governmental organisations (NGOs) and private Foundations
- bilateral aid agencies
- multi-lateral aid agencies
- development banks
- funds for development

A local NGO can easily turn to the funds of European or North American NGOs; the largest of these are represented locally by national offices or by their own representatives.

It is quite normal that the local NGOs contact other NGOs. Often European or North American NGOs manage government funds which are to be allocated to local NGOs. The European NGO will then act as an intermediary between his government and the local NGOs.

However, several NGOs have their own funds, which are collected from the people of their country, churches or lay organisations. These NGOs are accountable to their private donors. They therefore prefer to fund concrete projects: wells, mills, dams, agriculture and such, rather than training or costs of animation and institutional overheads.

They can be classified into the following categories:

a) Non-Governmental Organisations

- Protestant NGOs
- Catholic NGOs
- Muslim NGOs
- Buddhist NGOs
- Hindu NGOs
- Non-religious NGOs

b) Foundations

Large foundations such as Ford, Rockfeller, Aga Khan or others.

c) Embassy Funds

They are established in the capital of the country and managed by the Ambassador or his Officer-in-Charge of Development.

These funds can be used quickly without reference to the donor country. There is, however, a ceiling on these funds.

The following countries have active Embassy Funds:
Canada (CIDA), The Netherlands, Federal Republic of Germany (GTZ), U.S.A. (US-AID), Switzerland (DDA), Norway (NORAD), Sweden (SIDA), Finland (FINNIDA), etc.

Refer to Annexe I.5   *(a) Agreement between an Embassy and a Local Association to obtain finance for a Small Rural Development Project*

(b) Funds governed by Missions (Embassies) - approval documents of a project

d) Co-funding from the E.E.C. (European Economic Community)

This is not possible unless an NGO of a European country takes on the project and part finances it.

e) Multilateral aid from the UN (United Nations)

In particular from its specialised agencies such as UNICEF (children, youth, women), ILO (employment, technology, non-formal sector, handicrafts), UNESCO (literacy, education, communication), WHO (health and primary health care), UNIDO (small enterprises, industry) and UNEP (environment).

f) Bi-lateral aid from European/North American countries and their agencies such as:

- Canadian International Development Agency (CIDA) in Ottawa, Canada
- The Technical Co-operation of the Ministry of Foreign Affairs of the Netherlands Government in The Hague
- The GTZ (Deutsche Gesellschaft für Technische Zusammenarbeit) of the Federal Government of German Co-operation
- Mission d'Aide et Coopération (MAC) of the French Government in Paris
- US-AID in Washington
- Swiss Technical Co-operation (DDA) in Berne, Switzerland
- Government Agency of Development Co-operation of Belgian Government (AGCD)
- Development Co-operation Service of the Italian Government
- Norwegian Agency for International Development, (NORAD)
- Finnish International Development Agency (FINNIDA)
- (Swedish Development International Authority (SIDA)
- Danish International Development Agency (DANIDA), Ministry of Foreign Affairs, Copenhagen
- Overseas Development Ministry (ODM) in London, U.K.
- Japanese International Cooperation Agency
- Australian Development Assistance Bureau, Ministry of Foreign Affairs, Canberra, Australia
- Ministry of Foreign Afffairs, Technical Cooperation Service in Wellington, New Zealand

It is possible for an NGO to draw from bi-lateral aid grants, which consist of aid agreed between the government of the association's country and the donor's government. However, this is on condition that both parties give their support. Many rural groups have obtained such aid in larger amounts than the amounts disbursed through the respective local embassies.

g) Development Banks

The National Development Banks, Regional Banks, Asian/African Development Banks and the World Bank in Washington can, with the support of the local government, provide credit to rural groups.

h) <u>Other support</u>

Contact with a Ministry carrying out a government project covering the zone of activities in which the local Association works. Funds may be drawn from:

- small project programmes of EEC representation

- programmes of an UN agency: UNICEF, WHO, FAO, ILO

- bank loans from the World Bank, from the Asian/African Development Banks or the National Development Banks

<u>Some Advice</u>:

It is advisable to request support from a European or North American NGO which is working towards the same objectives as those of the local Association.

Refer to Annexe I.7 for useful addresses of organisations which fund NGOs

E. HOW TO CREATE FLEXIBLE OR REVOLVING FUNDS

The "project" approach for funding development activities is a system created by international funding organisations. It is a way of controlling/monitoring the funding process.

However, this system is not adaptable to the realities and needs of local development. Project-related aid finances only a part of their total needs. It establishes time schedules which are rarely followed, and often impose on local groups "administrative gymnastics" for which they are not prepared.

It is, thus, necessary to find another way to finance development.

If it is assumed that the donor wants, above all, partners in whom he has CONFIDENCE, meaning by that partners with a knowledge of planning, implementing and controlling their own management, it is best to find a new financing system which could avoid all the incumbent bureaucratic procedures and controls and which would allow an Association to manage the funds itself, according to its own requirements and priorities.

Flexible Funds

A development association should, therefore, negotiate with some of its donors to obtain flexible funds which are not allocated for specific activities.

Flexible funds are granted to well established organisations.

How to negotiate and obtain such funds

1. The Association should clearly define its policies and objectives and prepare a complete report on this subject (Refer Vol.I Part IV : Planning, Programming and Projects).
   The Association undertakes (provided it obtains the necessary means) to carry out a development programme and the donor undertakes to participate in its funding on a medium or long-term basis.

2. The report outlining the strategies will be a complete document comprising an accurate calculation of the total financial requirement needed for carrying out the programme.

3. The Association will call for funds which will be deposited at the beginning of each financial year, so as to facilitate the cash-flow.

4. The detailed allocation of funds received from the donor is carried out only by the Association and not by the donor. However, these funds should be utilised according to the agreement. An external

audit should be carried out by a Chartered Accountant who would certify to the donors how the money was spent.

5. Accounting is carried out after the expenses are incurred. The financial report submitted by the Financial Manager will enable each donor to find in the "income" column the exact amount of money granted to the Association and in the "expenditure" column, not only the expenses funded by his own contribution but all the expenses incurred for the general programme.

The donor usually finances a percentage of the Association's expenses. He will not call for any expense vouchers other than a progress report and an account of all the expenses.

Such a system obviously gives the Association a great deal of freedom of action. The funds thus received from donors can be just as easily allocated to a training programme or towards payment of a permanent animator's salary as towards meeting the cost of digging wells or constructing roads.

Refer to Annexe I.8: Technical Characteristics of Flexible Funds (Document given by B. Lecomte, SIX'S)

Revolving funds

This system is also interesting. One or several donors grant funds to a recognised association so that it could create its own development fund which it will manage by itself. The donor and the association sign an agreement which specifies the ways and means of estimating the funds required which will be regularly disbursed.

The association which receives these funds (as a grant) deposits the money in a bank. The funds are then distributed as loans to groups or to individuals according to specified rules.

These persons repay their loans to the Association which can then re-lend the money to other groups. The interest on these loans will help meet the cost of managing this fund, it will also progressively increase the capital.

Thus, the "Revolving" capital is regularly used by one group, then another and then more groups, and will help the association and its members to set in motion their own self-financing mechanism.

The system of revolving funds is, therefore, an improvement on the classical aid system.

Refer to Annexe I.9: Peasants of the Sahel originate a new international aid system.

F. PLANNING FOR THE END OF AID

It is useful and often necessary to receive external aid.
To receive such aid continuously, is to become dependent.
Planning an end to aid means becoming <u>independent</u> some day.

There are several stages of funding of a development programme:

a) Start out by depending on the Association itself (membership fees, own efforts, resources).

b) Add to this effort aid from a foreign donor to purchase machinery and equipment or to create revolving funds.

c) Substitute foreign aid (grants) with another form of aid i.e. credit (loans) at low interest.

d) Negotiate directly with a local bank which gives credit (at first with guarantees or security) to complete the self-financing of a programme.

A development programme which is directed towards independence (self-financing) should be ideally financed in the manner shown in the following example:

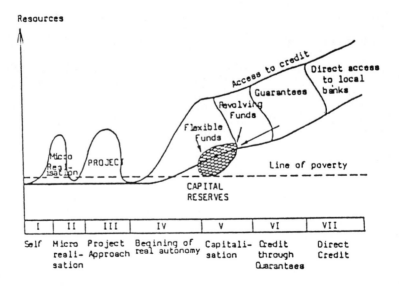

Independence cannot be achieved in a day.

Plan for the end of aid, the end of subsidized credit, as demonstrated in the above example.

Progress towards financial autonomy is grounded in the following elements:

From the beginning, the group:

- makes the maximum effort for its own development,

- obtains foreign aid for a specific purpose and for a specific period of time,

- manages its programmes efficiently,

- negotiates the required loans, when needed, for its development.

The most important instrument of financial autonomy is access to CREDIT.

He who speaks of credit spells SAVINGS.

How should an association organise a savings and credit system ?

N.B.: *Fernand Vincent and Piers Campbell will publish early 1989 another manual "Alternative Financing for NGOs and Development Associations". Write to the IRED Secretariat in Geneva if you are interested.*

PART TWO

SAVINGS AND CREDIT SYSTEMS

A. A LOCAL LEVEL SAVINGS SYSTEM

There is no local level development without LOCAL SAVINGS. Savings are a proof of a group's own efforts. Savings are the expression of a good management of income and expenditure in which the latter is reduced to a minimum.

Savings provide the key to investment and development. It is, therefore, essential to make the local groups save among themselves.

1. How to introduce the need to save

First educate the people and help them understand how to set aside a small amount of savings.

To educate is to train the people to realise that they do not necessarily have to be rich in order to save. No amount of savings is too small. Even a minimal effort is sufficient to set in motion the process of saving.

2. A chart showing the snowballing effect of a villager's savings

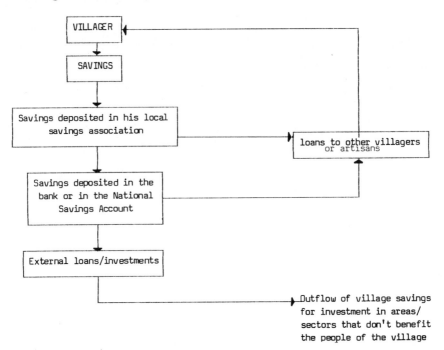

<u>Local savings</u> should be used for <u>local lending</u>.

Even if the peoples' savings end up in the city, it should be made to return to the village as loans to villagers who need credit.

3. Organising an internal savings system

Create a mechanism within the Association for collecting the savings of individuals and of groups for investing it safely to obtain the best returns.

To manage such a system, the following documents should be maintained:

- Register of Savings Account holders
- Individual Savings Accounts books
- Records of deposits and withdrawals of account holders
- Savings/Deposits bank account

a) The Association should maintain a <u>register of the savings account holders</u> of the Savings Bank.

SAVINGS ACCOUNT REGISTER

No.	Date	Surname/first name	Address	Misc. remarks
1.	10.5.86	Perera, Siripala	Wilpotha	
2.	11.5.86	Silva, Martin	Halawatha	
3.	11.5.86	Peries, Francis	Mundel	
4.	12.5.86	de Silva, Lionel	Angunuwila	Deceased 10.3.88
5.	etc.	etc.	etc.	
.	.	.	.	
.	.	.	.	

b) The Association maintains an individual savings booklet for each member of the Savings Bank where each deposit or withdrawal is recorded.

```
┌─────────────────────────────────────┐
│      Savings Bank of Kanakapura      │
│            SAVINGS BOOK              │
└─────────────────────────────────────┘
```

Name: Ravi Shekhar
Address: Kanakapura North, Handicrafts Association   in UC

Date	Description	Amount Deposited	Amount Withdrawn	Balance	Signature
10.1.1986	Savings deposit	50.00	–	50.00	of Cashier
25.2.1986	" "	50.00	–	100.00	"
31.3.1986	" "	100.00	–	200.00	"
30.4.1986	" "	25.00	–	225.00	"
15.5.1986	Withdrawn	–	100.00	125.00	of Customer

c) The Association should maintain its CASH AND BANK ACCOUNTS as illustrated:

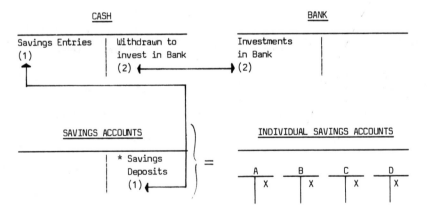

(1)   The day's or the week's savings (total of amounts deposited by the customers).

(2)   Date and amount invested in the bank to avoid theft and to earn interest.

4. What should be done with the savings of the Association ?

The savings of the Association are: amounts to the credit of those members who have deposited their personal savings; the profits from group projects (mills, rice bank, poultry, etc.); proceeds from collective labour or from amortisation.

What could be done with this money?

Most of the time, the money is deposited in a local commercial bank as a fixed deposit for a period of time and it earns, for example, 10% interest per annum.

What does the bank do with this money? The bank lends it to others (mostly non-villagers) for example at the rate of 15% interest per annum. The Association thus deposits its liquid resources in the bank while numerous villagers are in need of loans for development activities.

Negotiate with the bank so that the fixed deposits of the Association could be used as guarantee to those villagers requiring loans.

Thereby, the villagers "gain access" to credit which the bank has, up to now, refused.

But it can often happen that commercial banks refuse to lend to a group that has saved!!

As an alternative, an association can establish a Savings and Lending Bank. This bank will lend to those villagers in need on condition that all the members of the group hold themselves jointly and severally responsible in the event of default on repayment.

Note: For more information on the setting up of Savings and Credit Banks, write to:

- ASECA (Association of Savings and Credit Banks of Africa) P.O. Box 43278, Nairobi, Kenya and its National Associations

- Credit Union League of Thailand Ltd (CULT), 56/2 Moo 3, Ramkhamhaeng Road, Bangkok 10240, Thailand

- Self-Employed Women's Association (SEWA), Bhadra, Ahmedabad 38001, Gujerat, India.

B. ACCESS TO CREDIT

On several occasions, the need for development associations to gain access to local bank credit has been mentioned. Any productive project should be profitable. It is the norm to accept aid to start a project but this practice is not an advantage. An association should try and gain access to bank loans as soon as possible.

1. When to request a loan ?

An Association should not think of accessing credit until it has generated savings through its projects and gained experiences in the management of funds.

Credit is the complement of efforts made and is at the same time a result of such efforts.

Be cautious about borrowing money without being sure of repayment. The Association will be forced to sell all its hard earned goods in order to repay the loan.

2. Who gives loans ?

Traditional money-lender. These profit-making businessmen lend money at very high rates of interest. Avoid them.

The Commercial Banks. They are generally inaccessible to the villager. A bank will lend only to persons who have wealth and a capital which can be mortgaged (placed as a security) in case of default.

Development credit made available by the government: Agricultural Credit, Fisheries Credit, etc. To obtain these loans, too, some form of security is necessary.

The Savings and Credit Banks created by the villagers with their own savings in order to facilitate lending to other villagers who are members of an inter-dependent savings group.

3. How to manage a credit system ?

The savings bank created by the Association will make it possible to lend money to those who request loans.

a) The <u>Capital</u> can be made up of the savings of the members or it could consist of a foreign grant used as a revolving fund.

b) A <u>Credit Committee</u> (of 5 persons) can be appointed from within the Association to evaluate and decide to whom to lend and at what rate of interest.

c) In order to evaluate a loan application, the Committee would want to <u>ascertain if it can place confidence in the member requesting the loan</u>: they will have to ascertain whether he has worked well up to now and developed his projects successfully. The Committee would also like to know if the other members of his group agree to let him take a loan as they will be made jointly and severally responsible for the repayment of the loan together with him.

d) Once a decision is taken, the borrower should work out how to <u>repay</u> the loan and at what intervals.

The Association should consequently be organised to operate as a lending bank. The Treasurer should keep the following documents up-to-date:

i) <u>Promissory note/Personal Guarantee</u> which is the proof that a member has received a loan from the Savings and Credit Bank and that he undertakes to pay it back.

This document contains:

- the title "Promissory Note"
- the name and address of the borrower
- the name and address of the Savings and Credit Bank (or of the Association)
- the words "acknowledge owing ... to ..."
- the amount borrowed (in figures and in letters)
- contract to repay
- the dates and amounts of repayment instalments
- the interest rate
- the date, place and the signature of the borrower

Refer to Annexe II.1: Specimen Promissory Note

ii) The Account of each Borrower

LOAN ACCOUNT

LOAN GIVEN TO:

Surname, First Name :   PERERA, SIRIPALA
Address: Sapho Tani, Chiang Mai
Date of loan: 15th January 1988
Amount: 24,000 UC
Rate of Interest: 9 %
Repayment: in 2 years, every 3 months

in UC

Date	Details	Loan	Repayment Interest to be paid	Repayment Capital to be paid	Balance Capital
15.01.88	Loan at 9%	24,000.00	–	–	24,000.00
15.04.88	Repayment – Capital		–	3,000.00	21,000.00
	– 9% Interest		540.00		
15.07.88	Repayment – Capital		–	3,000.00	18,000.00
	– 9% Interest		472.50		
15.10.88	Repayment – Capital		–	3,000.00	15,000.00
	– 9% Interest		405.00		

NOTE:

- The monthly repayment is $\dfrac{24,000 \ UC}{24 \ Months} = 1,000 \ UC$

  Therefore, every 3 months 3,000 UC must be paid.

- The calculation of interest is as follows:
  9% per annum on 24,000 UC = 2,160 UC

  Therefore for 3 months $\dfrac{2160 \times 3}{12} = 540 \ UC$

  for the 2nd repayment the interest will be:

  9% per annum on 21,000 UC (24,000 - 3,000)

  then the interest for 3 months amounts to:
  $\dfrac{21,000 \times 9}{100} \times \dfrac{3}{12} = 472.50 \ UC$

iii) **A register of all loans granted** (which records all the loans grant-ed to members), to be maintained as follows:

in UC

Number	Date of joining	Date of loan request	Name and address of borrower	Loan requested		Loan granted		First instal-ment	Last instal-ment	Mode of pay-ment	Interest Rate
				amount	purpose	date	amount				
1	10.5.84	10.2.86	PERERA SIRIPALA WILPOTHA	15,000	GARDEN	31.3.86	12,000	30.4.86	30.4.87	CASH	10%
2	10.5.84	10.2.86	SILVA MARTIN MUNDEL	12,000	GARDEN	31.3.86	10,000	30.4.86	30.7.87	CASH	10%

iv) One copy of the loan repayment schedule will be attached to the Promissory Note/Personal Guarantee and another copy will be kept by the Treasurer and the Accountant.

The following document registers the exact conditions of repay-ment including the due dates on which instalments and interest are due :

REPAYMENT SCHEDULE

Borrower:
Loan: 12,000.00 UC on 1.04.86 for 1 year
Rate of Interest: 10% per annum
Repayment: Monthly

in UC

No.	Monthly instal-ment due on	Balance Amount Due	Instalment Due	Interest Due	TOTAL DUE	REMARKS
1	30 April 86	12,000.00	1,000.00	100.00	1,000.00	
2	31 May 86	11,000.00	1,000.00	91.66	1,091.66	
3	30 June 86	10,000.00	1,000.00	83.33	1,083.33	
4	30 July 86	9,000.00	1,000.00	75.00	1,075.00	
5	31 August 86	8,000.00	1,000.00	66.66	1,066.66	
6	30 September 86	7,000.00	1,000.00	58.33	1,058.33	
7	31 October 86	6,000.00	1,000.00	50.00	1,050.00	
8	30 November 86	5,000.00	1,000.00	41.66	1,041.66	
9	etc.					

PLACE AND DATE   READ AND APPROVED   TREASURER OF THE SAVINGS
                        THE BORROWER   AND CREDIT BANK

4. Interest

This is the "lending rate" of the bank.

The Bank pays a low rate of interest on savings and it lends at a higher rate. The difference is profit less management costs.

a) The rates of interest vary from case to case. Often, the Government helps villagers by allocating loans at subsidized rates of interest through Development Banks for specific activities; e.g. 4-6 % when the market rate is 14-18 % or more.

   This system is also followed by some donors who even go as far as not charging any interest at all.

   Even if such systems are acceptable on a short-term basis, it is advisable not to adopt them on a long or medium-term. This is a disguised subsidy which will strain relations between villagers as some will have access to advantages for which the others are not eligible.

b) The calculation of interest

   If the interest rate is fixed, (for example at 10 %), it means that after one year the borrower will pay 10 % of the borrowed amount plus the depreciation of the debt.

   Another example is given below:

   Calculation of interest on a loan of 60,000 UC at 8 % per annum.
   Total interest for the year: $60,000 \times \dfrac{8}{100} = 4,800$ UC

   If the loan is paid back in 2 years in 2 instalments, at the end of the 1st year Capital repayment should be:
   $$\dfrac{60,000}{2} = 30,000 \text{ UC}$$

   Hence, the yearly repayment

   = 30,000 (capital)
   + 4,800 (interest)

   = 34,800 UC

If the loan is to be paid back monthly:

$$\text{(a)} \quad \text{Capital repayment} = \frac{60,000}{24} \qquad = 2,500$$

$$\text{(b)} \quad \text{Monthly interest} = \frac{60,000 \times 8}{100} \times \frac{1}{12} = 400$$

$$\text{Total repayment per month} = \underline{\underline{2,900}}$$

Therefore, the 1st monthly instalment to be repaid
= 2,500 + 400 = 2,900 UC

However, the 2nd month's interest will be a little less than the 1st month's since a part of the Capital has already been paid back.

(a) Balance Capital due = 60,000 - 2,500   =   57,500 UC

$$\text{(b) Interest} \qquad = \frac{57,500 \times 8}{100} \times \frac{1}{12} \qquad = \quad 383.33 \text{ UC}$$

(c) Therefore, the 2nd monthly instalment plus interest will be:

$$2,500 \text{ (Capital)} + 383.33 \text{ (Interest)} = \underline{\underline{2,883.33}} \text{ UC}$$

and so on, until the whole debt has been repaid.

C. COLLATERAL / GUARANTEE

When granting a loan, the bank will require the mortgage of an asset as collateral to ensure recovery in the case of non-repayment. If no such assets are available, then a guarantee should be submitted as security.

A guarantee is a form of security given on behalf of the borrower to the bank by a solvent person or institution as a commitment to repay the debt of the borrower in case of default.

Refer to Annexe II.2: A Specimen Guarantee

An International Foundation called RAFAD (Research and Application for Alternative Financing for Development), Case 117, 3 rue de Varembé, 1211 Geneva 20, Switzerland, recently established a system of offering guarantee funds (FADEP) in favour of local groups who have already accumulated their own savings, proved their responsibility and professional competence and who need a guarantee to obtain credit from their local banks for their development projects.

Refer to Annexe II.3: Application to obtain a RAFAD Bank Guarantee

An association can contact them to make use of such guarantees but before doing so it should make sure that its local bank will accept a letter of credit from a recognised European Bank as a guarantee and not insist on a physical transfer of funds.

Similar schemes are operated by the Women's World Bank in New York (104 East 40th Street, 6th Floor, New York, N.Y. 10016, USA) and friends of the Women's World Banking Association in Thailand (c/o Bangkok Bank Ltd., 333 Silong Road, Room 150911, 15th Floor, Bangkok 10500).

The system of Savings and Credit Banks works on the principle of offering security 'jointly and severally.' This means that in an interdependent group, when one person borrows, the others are jointly and severally liable and financially responsible for his loan repayment. If the borrower does not pay, the other members of the group will have to repay the entire debt.

Unless the borrower can be trusted, responsible persons should not sign a security or guarantee on his behalf.

PART THREE

SIMPLE ACCOUNTS, BUDGETS AND CASH-FLOWS

A. THE ACCOUNTING SYSTEM

The proper management of a local development association depends largely on the efficient organisation of its accounting system and finances.

This chapter is of great importance and has a number of practical documents which will prove useful in the day to day accounting operations.

Accounting helps to:

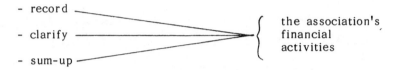

- record
- clarify   the association's financial activities
- sum-up

Its purpose is to provide information regularly regarding the Association's financial position.

The accounts should, therefore, always be up-to-date. In small associations the entries will be made daily or at least weekly.

A simple accounting system

1. Every petty-cash operation (mill, poultry, rice bank) is recorded in an exercise book according to a method described later on.

2. The Treasurer and another literate person maintain a box file to keep all accounting documents including the above-mentioned exercise books.

3. Each accounting document is numbered in correspondence with the cash book or bank book: the same numbers are written against the corresponding exercise book.

4. If there is no supporting voucher, the treasurer prepares one by using a blank paper on which he writes the maximum details concerning the expense incurred, giving the reasons why a voucher could not be obtained.

5. The voucher file as well as the cash and bank books are kept in a secure place in a cupboard in the Secretariat premises.

If each such document is filed immediately so that it will not be lost, and if this procedure is strictly observed, it will enable any qualified accounts clerk to write out the annual accounts of a local association in a few days.

B. ACCOUNTING DOCUMENTS

A simple accounting system uses the following documents:

<u>CASH</u>

- withdrawal vouchers/receipt
- deposit vouchers
- cash receipts which are prepared when there is no supporting document
- currency verification forms

<u>BANK</u>

- bearer cheques
- cheques not crossed
- crossed cheques
- withdrawal forms - debit/credit advice
- signature cards
- bank statements

<u>OTHER ACCOUNTING DOCUMENTS</u>

- invoices
- stock entry forms
- petty cash vouchers
- etc.

Refer to Annexes III.1 - Cash account vouchers

III.2 - Bank account vouchers

III.3 - "Goods" account vouchers

C. KEEPING THE BOOKS OF ACCOUNTS

The accounts are maintained in accounts books which record the financial <u>operations</u> (accounts entries). These should always be <u>supported by accounting documents</u>.

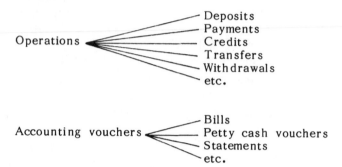

The CASH Book

Use an exercise book and write on the front cover:

- the name of the Association
- in capitals: CASH
- the year

Keep the accounts in this book, day by day, month by month (one/two pages per month) according to the example below:

CASH BOOK

in UC

Date	Document	Details of operation	IN	OUT	BALANCE
1988					
1.1.88	1	Transfer	17.00		17.00
1.1.88	2	Repair heater		4.00	13.00
3.1.88	3	Fuel motorcycle		1.75	11.25
3.1.88	4	Purchase Mill Oil		6.00	5.25
9.1.88	5	Withdrawal bank	20.00		25.25

The balance at the end of the day can be seen by referring to the book. The cashier has only to reconcile this balance and the actual cash-in-hand by using the "currency inventory" form (refer to Annexe VII.2) to know the exact cash-box balance.

The cash must be deposited in a lockable cash box and kept in a place secure from any possible theft.

The BANK Book

When the need arises, open a Bank Account to deposit funds - refer to Annexe II.2 (a). The signatories of the bank account should be members of the Executive Committee. The signatures of the Chairman, Treasurer and Secretary should be submitted to the Bank. Any two should sign cheques to authorise withdrawals/transfers.

BANK BOOK

COMMERCIAL BANK

Account No. 333.215.20.A
Rural Development Society, Sibu   in UC

Date	No.	Details	IN	OUT	BALANCE
1988					
1 Jan.		Opening balance			11,000
5 Jan.	1	Received from CIDA	45,000		56,000
8 Jan.	2	Purchase of water pump		3,500	52,500
15 Jan.	3	Purchase of bicycle		2,500	50,000
30 Jan.	4	Withdrawal for petty cash (Cheque no. 342570)		6,000	44,000

All documents should be carefully numbered and filed in a BANK file. Each document number will correspond to the relevant entry in the account.

All foreign grants and large sums received should be deposited in the bank and recorded in the bank book. The cash box is for petty cash. The money needed as petty cash will be withdrawn from the bank.

Regularly reconcile the records with the bank statements and check if there are any discrepancies.

Do not forget to record interest payments and bank charges before calculating the final balance. Refer to the bank statement for the exact amounts.

D. THE BUDGET

1. The connection between the budget and the programmes of an association

The budget is defined in financial terms as "an expression of a plan of a programme", and can be diagramatically shown as :

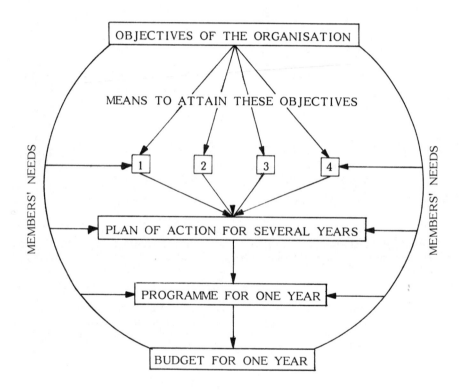

Usually the Constitution of an Organisation defines not only its objectives but also the means by which it hopes to achieve these objectives.

---

* B. Tiotsop, IRED Management Seminar, Ouagadougou, Burkina Faso, May 1985

These objectives should be achieved according to a plan over a period of several years; taking into account expected financing. The long-term plan should be divided into shorter periods of time, for instance, one year. The short-term plans are called programmes which cannot be worked out without a budget. Planning should reflect the needs of the association and of the entire community.

The budget can be broken into several parts, e.g.

- Investment (if the organisation has production activities)

- Equipment (as distinct from "investment", which is directly profitable - heavy expenses on equipment can be harmful to an organisation)

- Various operations which generate an income for the organisation

- Administrative and overhead costs which are difficult to apportion among all the different activities.

2. The budgetary functions

The various budgetary functions can be summarised as follows:

- Once the programme is worked out, it is necessary to reflect on the means to implement it. One of the primary functions of the budget is the MOBILISATION OF RESOURCES FOR THE REALISATION OF THE PLANS AND PROGRAMMES.

- The budget is also the means through which the available resources are apportioned. It enables the association, therefore, to decide on the amounts required for each programme; reducing some expenses and increasing others.

- The budget is a protective instrument. It determines both the limits that should not be exceeded where expenses are concerned and the minimum amount that should be earned as income.

The budget is an indispensable instrument which no individual, family or organisation can do without. The more numerous the activities, the larger the means and more complex the budget.

3. Preparing a budget

What has been described so far is budgeting under "ideal conditions": i.e. where a programme is first structured to meet the felt needs of the members and the community and a budget is then prepared to support the programme.

In practice, however, budgets do not evolve in that way. More often an association gets funded through individual projects which do not generally evolve through the steps recommended in the ideal model. Projects adjust themselves to the needs of an association or a community in the course of implementation. They are fairly rigidly managed against an estimate of income and expenditure.

An organisation whose activities are funded through individual projects should incorporate individual project budgets into a single consolidated programme budget and perceive each project as a part of a single consolidated programme.

The preparation of a "Programme budget" commences with the income and expenditures estimates of the different parts of the programme. Some parts of a programme will earn an income which can be used for financing other parts of the programme.

A budget is a type of "forecast". Three problems that arise in the course of such forecasting are :

- not accounting for large amounts of expenditures or income due to inexperience: not having sufficient experience in this field, may be a very real problem. The problems can be solved or minimised by finding out from other organisations involved in similar operations the accounting methods to adopt or by using a detailed accounting plan, etc.

- the need to have accurate reliable data. The best information sources are the accounts of the previous years on which forecasts can be confidently based. In the alternative, one has to rely on the experience of others, which may not be applicable for one's own association since each organisation has its own style of functioning.

- the link between estimates and efficiency. On the one hand, there is the risk of spending a great deal of money to earn a meagre income. On the other hand, for the same amount of work, the cost can vary largely, depending on how efficiently the work is carried out.

4. Some examples

Every association should work out its annual budget, and include in it each and every activity and project which has been planned.

To work out a budget, start with the income expected for the coming year. It is not easy to estimate this income.

If it is really impossible to estimate the income, even approximately, then the Management of the Association will have to be on a "day-to-day" system as a budget calculation will not make any sense.

However, if, for example, the association knows in advance that two donor agencies are going to grant certain funds, and that a certain amount could be collected as membership fees, then it is in a position to evaluate the income and consequently estimate the expenses.

The budgetary income and expenditure can be estimated accurately if sufficient effort is expended.

The income and expenditure of a local Developement Association can be presented under the following "headings" and "sub-headings":

Income

- The Association's own income
- Income from donors

Administrative costs

- Salaries, allowances and social expenses of the leaders and animators
- Travelling expenses and support costs of programmes
- Communication costs
- etc.

Programme budgets

- income/expenditure of project A (rice bank)
- income/expenditure of project B (mill)
- income/expenditure of project C (wells)

(See details in following examples)

5. The annual budget

As an example, the Annual budget of a village rural development association for the year 1988 is given below:

VILLAGE RURAL DEVELOPMENT ASSOCIATION

Budget 1988

Estimated Expenditure	UC	Estimated income	UC
Animators allowance	48,000	Canadian Embassy grant	60,000
Administrataive costs	20,000	Oxfam grant	30,000
Poultry costs	15,000	Membership fees	25,000
Purchase of Mill	100,000	Income from Mill	140,000
Cost of running Mill		Poultry income	20,000
(oil, gasoline)	52,000		
	235,000		
Profit	40,000		
	275,000		275,000

To prepare budgets for a specific programme, take into account the following items:

A training course: budget items: participants' travelling, particpants' food and lodging, lecturers' fees, field work, cost of publication of report...

A health programme: animators' and nurses' salaries, purchase of medicines, transport costs, administrative costs...

A rural development programme: animators' salaries, cost of seedlings, plants, transport, advertising costs, general and administrative costs... etc.

A programme for starting a enterprise: directors' and workers' salaries and fees, purchase of raw material, inputs, general and administrative expenses...

The following example is more complex:

6. A 3-year budget for an activity of an association

The programmes of the Association are:
- cultivation of collective fields
- operating a mill
- carrying out training programmes
- running a health clinic
- digging wells

The Association plans to purchase a second hand mill (at a cost of 140,000 units of currency) to increase the milling capacity of the village but does not have the necessary funds.

It has its own resources, but in order to develop, the Association is counting on receiving foreign aid.

How will it estimate the total funds necessary?

1. It prepares an annual budget for each activity:

(i)   Budget for cultivating the "Collective fields"

Expenses	UC	Income	UC
Allowances	20,000	Sale of produce	125,000
Inputs, pesticide	10,000		
Fertilizer	15,000		
Seedlings/plants	10,000		
Transport	12,000		
Miscellaneous	8,000		
Total expenses	75,000		
Estimated profit	50,000		
	125,000		125,000
	=======		=======

(ii)   Budget for the Mill Operation

Expenses	UC	Income	UC
Gas/Oil	28,000	Milling income	100,000
Supplies	6,000	(12 months	
Transport	7,000	x 66700 kg	
Miller/Mechanic		x 1.5 UC)	
labour charges	28,000		
Miscellaneous	1,000		
Total expenses	70,000		
Depreciation	30,000		
	100,000		100,000
	=======		=======

(iii) Budget for the Training Programme

Expenses	UC	Income	UC
Teachers' allowances	10,000	Fees	
Participants' food	12,000	50 x 20	1,000
Travel	4,000	Net cost	29,000
Books	4,000		
	30,000		30,000
	======		======

(iv) Budget for the Health Clinic

Expenses	UC	Income	UC
Nurses' allowance	15,000	Patients' fees	10,000
Purchase of medicines	15,000	Net cost	25,000
Transport & Misc.	5,000		
	35,000		35,000
	======		======

(v) Budget for "Coordination"

Expenses	UC	Income	UC
Coordinators'/Animators'		Fees	
Allowances	20,000	100 x 50	5,000
Travel	8,000	Net cost	35,000
Supplies, postage	8,000		
Miscellaneous	4,000		
	40,000		40,000
	======		======

2. The Association summarises the estimated income and expenditure for the year 1987

SUMMARY OF EXPENSES AND INCOME

	EXPENSES	INCOME
1. Collective fields	75,000	125,000
2. Mill	100,000	100,000
3. Training	30,000	1,000
4. Health clinic	35,000	10,000
5. Coordination	40,000	5,000
Total expenses	280,000	
Total income		241,000
Balance needed		39,000
	280,000	280,000
	=======	=======

In 1987, the Association's budget will, therefore, be:

Total expenses	280,000
Income	241,000
Balance necessary to run present operations	39,000
Purchase of 2nd Mill to operate in 1988	140,000
Funding needed	179,000
	=======

The 1988 budgets will be identical to those of 1987 but no invest-ment will be made. However, the cost of the operating the 2nd Mill will be added to the budget this year.

In 1989, the Association expects that the income from the collective fields will increase to 150,000 UC and that the income from the mills will increase to 120,000 UC each. However on the expenditure side, the costs of the Health programme and the cost of coordination will increase to 40,000 UC and 45,000 UC respectively.

SUMMARY

BUDGET OF THE ASSOCIATION FOR 1987, 1988 AND 1989:

EXPENSES	1987	1988	1989	TOTAL
1. Collective fields	75,000	75,000	105,000	255,000
2. Mills	100,000	200,000	200,000	500,000
3. Training	30,000	30,000	30,000	90,000
4. Health	35,000	35,000	35,000	105,000
5. Coordination	40,000	40,000	45,000	125,000
	280,000	380,000	415,000	1075,000
Investments				
Second hand Mill Bullock carts	140,000		35,000	175,000
TOTAL EXPENSES	420,000	380,000	450,000	1250,000
INCOME				
1. Collective fields	125,000	125,000	150,000	400,000
2. Mills	100,000	200,000	240,000	540,000
3. Training	1,000	1,000	1,000	3,000
4. Health	10,000	10,000	10,000	30,000
5. Coordination	5,000	5,000	5,000	15,000
Income	241,000	341,000	406,000	988,000
Balance needed (foreign aid)	179,000	39,000	44,000	262,000
TOTAL INCOME	420,000	380,000	450,000	1250,000

To summarise, the budgetary position of the association as estimated for the next 3 years :

	UC
Budget expenses 1987-1989	1,250,000
Estimated income 1987-1989	988,000
Funds to be found	262,000

Donors

The above financial estimates supported with an introduction of the Association and its proposed activities should be forwarded to the donors.

Two donors, for instance, could finance the programme:

Donor A (CIDA-Canada for example)	112,000 uc
Donor B (NORAD for example)	150,000 uc
Total	262,000 uc
	=======

> If funds from donors are needed in January 1987, send the dossier by March 1986!

E. CASH-FLOW

The Cash-flow of an association consists of the inflows and outflows of the money in the cash box and in the various bank accounts.

THERE IS A CLOSE RELATIONSHIP BETWEEN THE BUDGET AND THE CASH-FLOW.

If the programmes and projects are to be carrried out as planned, then the necessary funds should be available at the needed time to pay for the purchase of materials, transport of produce, salaries, etc.

To avoid problems, the cash-flow of an association should be estimated every month so that it could be decided how much then should be in the cash-box as "petty cash" and how much then should be in the bank account to meet expenses.

1. The cash-flow plan

To propose a cash-flow plan, it is thus necessary to break down the annual budget into monthly budgets. An example of a cash-flow plan of a development-oriented organisation follows.

CASH-FLOW PLAN
of a development-oriented village organisation

MONTH	JAN.	FEB.	MARCH	APRIL	MAY	JUNE	JULY	AUG.	SEPT.	OCT.	NOV.	DEC.
OUTFLOWS:												
Salaries	15,000	15,000	15,000	15,000	15,000	20,000	20,000	20,000	20,000	20,000	20,000	20,000
Communication costs	3,000	2,500	1,000	2,500	3,000	2,500	3,000	1,000	2,500	3,000	3,000	2,500
Administration costs	5,000	5,000	5,000	5,000	5,000	5,000	5,000	5,000	5,000	5,00	5,000	5,000
Granary	10,000	8,000	5,000	9,000	10,000	8,000	7,000	6,000	7,500	8,000	6,000	7,500
Mills projects	15,000	13,500	15,000	16,500	15,000	12,500	13,000	12,000	14,000	12,000	11,000	12,000
Total expenses	48,000	44,000	41,000	48,000	48,000	48,000	48,000	44,000	49,000	48,000	45,000	47,000
INFLOWS:												
Opening cash/bank Balance	10,000	97,000	71,000	50,000	24,000	NIL	127,000	101,000	77,000	48,000	20,000	40,000
Own funds	15,000	18,000	20,000	22,000	24,000	25,000	22,000	20,000	20,000	20,000	15,000	15,000
Grants : - CIDA	120,000	–	–	–	–	–	–	–	–	–	–	–
- NOVIB	–	–	–	–	–	150,000	–	–	–	–	50,000	–
Total inflows	145,000	115,000	91,000	72,000	48,000	175,000	149,000	121,000	97,000	6 9,000	85,000	55,000
CASHFLOW	97,000	71,000	50,000	24,000	NIL	127,000	101,000	77,000	48,000	20,000	40,000	8,000

The Association will not have any cash-flow problems. It can meet all its commitments since the required funds will be available in the cash-box and in the bank account of the association at the beginning of each month.

According to the cash flow plan, two donors will be transferring their grants to the bank account in January, June and November: The Association will face serious cash-flow problems if these transfers get delayed for some reason or other. The Association should plan its cash-flow to meet such a situation as well.

2. Investing available funds

An Association should not leave too much money in its current account in the bank.

If the cash-flows have been carefully worked out the amounts that had to be withdrawn from the bank and the times of withdrawal would be known in advance.

Do not let the money stay idle in the bank. Invest it!

In the short-term, (one, three or six months), the available liquid assets should be invested in a fixed deposit. The cash-flow plan would indicate the amount of liquid assets available for short-term invest-ment.

The interest on a current account is nil or almost nil. The interest on a fixed deposit is 8-14 % or more according to the amount and the country.

If the cash available in the bank is in reserve funds that will not be utilised for two or three years, then the capital can be invested.

The Association should seek advice from the bank and not take undue risks. Short term investments of idle cash can increase the cash turnover by 2, 3 or 5 percent.

An Association can use its reserves to give guarantees/securities to its members who may apply for bank loans. But first a decision of the General Assembly or the Executive Committee must be made regarding the use of reserve funds for such guarantees.

F. STOCK MANAGEMENT

The stock represents the quantity of products remaining in storage. This includes grain, fertiliser, insecticide, etc.

At the end of each financial year, it is very important to make an inventory of the various stocks.

Stock-keeping entries are similar to cash-book entries in that quantity-wise entries are made and the increase and decrease of the different stock items calculated and recorded.

STOCK CARD
RICE BANK
Rural Development Association, Chiang Mai

Date	Description	IN (kg)	OUT (kg)	BALANCE (kg)
1.1.88	52 bags of rice of 50 kg	2,600		2,600
3.1.88	9 bags - Village produce	450		3,050
4.1.88	12 bags Chiang Mai	600		3,650
9.1.88	26 bags - sold to Chiang Mai Consumers' Co-operative		1,300	2,350

The Stock Account

Keep stock cards for all the products which will not be sold immediately. The above example is of a stock account of a rice bank that buys rice in large quantities and sells to villagers through a local co-operative at reasonable prices.

To fix the selling price of a product to members of the Association, add to the purchase price the cost of transport and storage and the cost of managment.

Keep an account of the stocks of each item, separately:
- fertilizer
- watering cans and equipment for the group's cultivation plots
- petrol and grease
- spare parts for milling equipment
- etc.

The following documents should be maintained for proper stock management:

- a stock card for each product
- a purchase order book and a goods-received book
- a specimen invoice for use by the Association
- a stock account

Stocks can be maintained in kilos, litres, number of pieces and/or values. Values will be calculated as per purchase price of product, or if it is a product that can be stocked (such as cereal), as per minimum selling price.

The Stock-keeper is responsible for the stocks. For efficient management of stocks he should maintain:

- STOCK ENTRY VOUCHERS to record members' deposits (each voucher should have a duplicate carbon copy - the original being for the member).

```
                            GRANARY
                  YOUNG FARMERS ASSOCIATION
                    Stock Entry Voucher

We acknowledge the receipt of :
                  5 bags of rice of 50 kg each (250 kg)
from : Mrs. Karunawathie Menike
of :   Wilpotha Womens' Handicraft Association
       Wilpotha
on :    17th September 1988
                                     Pieris
                            STOCK-KEEPER
                            (F. Peries)
```

- an OUT VOUCHER BOOK to record members' withdrawals (in duplicate)

```
                            GRANARY
                  YOUNG FARMERS ASSOCIATION

                    STOCK OUT VOUCHER

I, (name)....................................., acknowledge the receipt of:

2 bags of rice of 50 kgs each (one hundred kilos)

from the stock-keeper: Mr. H. Fernando
on: 17th September 1988.


                                     _____
                                     Signature of member
```

- ## STOCK RECORDS BY PRODUCT

YOUNG FARMERS ASSOCIATION
STOCK CARD - GAZ/OIL

Date	Description	IN	OUT	BALANCE
1988				
1 Jan.	in stock	150 lt.		150 lt.
5 Jan.	Toyota van		50 lt.	100 lt.
6 Jan.	mill...		30 lt.	70 lt.
10 Jan.	purchased at TEXACO	500 lt.		570 lt.
11 Jan.	mill...		50 lt.	520 lt.

ACCOUNTS PER MEMBER can also be kept (e.g. rice bank)

G. MANAGEMENT OF VEHICLES

It is possible that when the Association expands, a donor finances the purchase of a car or a van that may be necessary for the smooth running of a particular project or programme..

Note: Think well before acquiring a vehicle. Its maintenance is very costly and often weighs heavily on the management by eating into profits.

If a vehicle is purchased, manage it efficiently by taking the following steps:

Open a file for all the documents pertaining to the vehicle, including the set or rules for its proper use, the maintenance manual, etc...

Decide how to depreciate the vehicle, i.e. how to spread, over a period of 4 years, the purchase price of the vehicle. If it is estimated that the life of the vehicle is around 100,000 km, it will be advisable to sell it after that. If it is expected to run 100,000 km in 4 years, then it will run about 25,000 km in any one year. It is on this basis that the cost per km should be calculated.

To calculate this cost enter all the expenses relating to the consumption of fuel, maintenance, repairs and make a calculation on the following lines :

COST PER KM CALCULATION (TOYOTA HIACE VAN)

	Purchase price
÷	4 years (lifetime of vehicle)
=	X
+	fuel and oil costs per year
+	insurance, taxes, license fees per year
+	maintenance & repair costs per year
=	TOTAL per year
÷	number of km run per year
=	COST PER Km
	==============

Refer to   Annexe III.4: Example of km cost calculation for a Toyota van

Making such a calculation will show that a vehicle is very costly to operate and maintain.

The vehicle should be in the charge of one person. The driver should be responsible for the vehicle. The driver should maintain an up-to-date running chart of all the day-to-day trips according to the following format:

RUNNING CHART OF VEHICLE

Van - Toyota Hiace
22 Sri 6437

Date	Kilometre reading at departure	Kilometre reading on returning	No of km run	Reason for use	Signature of driver
8.1.88	12,102	12,182	80	Purchasing seeds	Banda
9.1.88	12,182	12,200	18	Attend General Assembly at Mundel	Sena

Each run should be entered separately in this register and signed by the person who used the vehicle. He should see to it that the register is kept in the vehicle. Tie a ball-point pen to the register. Fuel consumption, too, should be recorded.

FUEL CONSUMPTION REGISTER

Van - Toyota Hiace
22 Sri 6437

Date	Place of purchase of fuel	Km reading	No. of litres	Purchase price	Remarks
16.7.88	Halawatha	11,650	40	13.50	Cash purchase
20.7.88	Colombo	13,540	40	13.50	On account

From time to time calculate the fuel consumption per 100 km, as for example:

$$\frac{\text{No. of litres consumed}}{\text{No. of km run}} = \frac{60 \text{ lt.}}{412 \text{ km}} = 14.5 \text{ lt. for 100 km}$$

The fuel cost per km can be calculated, as for example:

$$\frac{\text{Consumption for 100 km X fuel price}}{100} = \frac{14.5 \text{ X } 13.50}{100} = 1.95 \text{ uc per km}$$

As explained earlier, it is also advisable to keep a maintenance book for the vehicle in the office to record daily expenses.

VEHICLE MAINTENANCE BOOK

Van - Toyota Hiace
22 Sri 6437

Purchased on 12th June 1988 from United Motors Ltd., Colombo
Purchase price = 175,000 UC

in UC

Date	Work carried out	Place Garage	Invoice Date	Amount	Signature
16.8.88	Cleaning, greasing	Havelock Town	18.7.88	100	Sena
20.9.88	Clutch repair	Reid Avenue	29.7.88	350	Sena

Enter the Insurance details in the same book.

The vehicle costs are incurred in respect of various projects and programmes. It is, therefore, correct to apportion to each project or programme the corresponding charges for the use of the vehicle. Hence, each month the total distance run can be divided among these activities.

FORM FOR APPORTIONING VEHICLE USAGE AMONG VARIOUS PROJECTS

Month	No. of km run	Poultry Project	Livestock Project	Granary Project	Rice Mill	Health Project	Collective Field	Misc.
June	350	25	46	30	30	120	50	49
July	322	21	40	30	30	110	46	45

If the cost per km is calculated, it will be easy to calculate the charges to be made against each project each month;

e.g. in <u>June</u>

<u>Poultry Project</u>: 25 km X 1.95 uc/km = 48.75 UC

<u>Livestock Project</u>: 46 km X 1.95 uc/km = 89.70 UC

PART FOUR

MANAGEMENT
OF
SMALL PROJECTS

A. Cereal Bank

B. Rice Mill

A Development-oriented Association should regulary undertake the management of <u>projects</u> which benefit the Community. Here are some examples of such projects:

- a cereal bank
- a mill
- a collective cultivation plot
- a village health clinic
- a training programme
- a water management project

The efficient management of a small project will depend on:

- <u>office management</u> - management committee, regulations, secretariat, administration
- <u>technical management</u> - techniques to be used, inputs, machinery and equipment
- <u>personnel management</u> - organisation of collective work, labourers, motivation, leadership
- <u>stock management</u> - costs, fixing selling or buying price, depreciation, budgets

Here are three detailed examples of the management needs of small projects:

A. MANAGEMENT OF A CEREAL BANK

1. Administrative management

(a) <u>Formation of a Management Committee</u> (e.g. 6 men and 6 women) to have a fair representation and to mobilize people to participate actively.

The Committee divides among themselves the responsibilities: Chairman, Secretary, Treasurer, Auditors, etc.

(b) At every meeting of the Committee of Management, the Secretary records all the decisions in a <u>Book of Minutes</u>.

(c) The Committee adopts <u>rules of management</u> for the cereal bank and compiles a <u>list of all those who become members</u>.

<u>The Committee makes all the important decisions</u> concerning the construction of buildings and purchase of grain and fixing of the selling price.

It also designates the stock-keeper and determines his remuneration. The stock-keeper is the cereal bank's outside representative.

(d) <u>The selling price of cereal is fixed by the Committee</u>

It should be below the market price and it should conform with the price levels fixed by the government. The price can vary from one region to another depending on the supply and demand position in each region.

2. Technical management

It will be necessary to construct buildings for the cereal bank. Obtain a cost estimate and ask for a grant or a loan from a foreign donor.

The Committee should open a project file for the construction of the cereal bank and work out an income and expenditure budget.

The Committee should decide through whom the bank will be constructed and should organise the participation of the people in the construction.

As soon as the premises are ready, the Committee should employ a stock-keeper. Then the project enters its operational phase.

When storing cereals, supervise the stocks and treat the grain. Sprinkle kitchen salt on the floor and place the cereal bags on planks; each type of cereal should be placed in a separate place within the premises.

3. Personnel management

The Committee should employ a stock-keeper, who will be the only person to receive an allowance, unless his services are voluntary. The stock-keeper manages the stocks, sells the cereal, keeps the accounts and maintains the premises.

4. Management of stocks

RICE STOCK BOOK

CEREAL BANK

DATE	OPERATION	IN	OUT	BALANCE	REMARKS
20.4.88	Purchase of rice	300 bags		300 bags	
21.4.88	Daily sales (see below)		5 bags	295 bags	

The column "remarks" serves to register "opened" or "lost" bags.

In principle, the stock-keeper pays cash for the purchase of cereals, or else he will sign a "Goods Delivered Note" from the supplier and will await an "invoice" before paying through the bank.

The stock-keeper's sales are registered daily in a special book and totalled each month.

RICE SALES BOOK

CEREAL BANK

in UC

DATE	OPERATION	NO. OF BAGS SOLD	UNIT PRICE	TOTAL PRICE	AMOUNTS COLLECTED	BALANCE TO BE PAID
21.4.88	To Bimo	2	11.00	22.00	22.00	-
21.4.88	To Ismail	3	11.00	33.00	25.00	8.00
30.4.88	TOTAL	5	-	55.00	47.00	8.00

To simplify matters, the last column will be used to register the "balances due". When the person pays the balance due from him, it is entered in red ink.

As soon as the stock-keeper has collected a certain amount of cash, (the limit being determined by the Management Committee) he should take it to the bank himself, or else to the Central Cashier of the Association who will deposit this money in the bank. The stock-keeper should not keep too much cash with him due to risk of theft.

If it is the stock-keeper's responsibility to deposit cash in the bank, a bank deposit register should be maintained by him:

BANK DEPOSIT REGISTER

DATE	DETAILS	AMOUNT	BALANCE IN A/C
22.4.88	Deposited - Receipt No: 127234	500.00	500.00
25.4.88	Deposited - Receipt No: 130031	250.00	750.00
30.4.88	Deposited - Receipt No: 147859	300.00	1050.00

The Management Committee (or the Advisory Committee of the Association) would have named the signatories under whose signatures cash withdrawals could be made from the savings bank to purchase stock or to fund another rice bank.

It is best that all withdrawals or cheques be signed by two signatories; one designated by the local Management Committee and the other by one of those in charge of the Association, such as the Executive Secretary.

The authosised signatories should give their specimen signatures to the bank and should derive their authority from the Administrative Committee of the Association which took the decision to open the bank account in the name of the Association.

Note: An Association should never open an account in the name of an individual.

5. Accounts management

The Association should find out at the end of each year whether the cereal bank was run at a profit or at a loss.

All the expenses and the income for the year should be accounted for.

(a) Income

Calculating total income is simple. Add the total of income/ receipts for each month by referring to the sales register.

Also add all the other income such as the bank interest received on the deposits.

(b) Expenses

The main cost items of a cereal bank will be the following:

- the purchase of cereals
- transport costs related to purchasing
- loss of grain
- allowance paid to stock-keeper
- maintenance cost of premises
- cost of preservatives - salt
- cost of insecticides
- depreciation of premises
- sundry expenses

(c) Depreciation

The depreciation of the premises can be calculated over, say, 20 years or more.

Divide the cost of construction of the building by the number of years to arrive at the amount of depreciation per year.

$$\frac{160,000 \ UC}{20 \ years} = \frac{Construction \ cost}{depreciation \ period} = 8,000 \ UC/year$$

An example of the final accounting for the year of a cereal bank is the following :

THE PROFIT AND LOSS ACCOUNT
OF THE CEREAL BANK
FOR THE 12 MONTHS ENDING 31 MARCH 1988

EXPENSES/COSTS	UC	INCOME	UC
Purchase of cereals	125,000	Sales of cereals	165,000
Transport	6,000	Bank interest	15,000
Loss of grain	2,000		
Stock-keeper's allowance	10,000		
Maintenance	4,000		
Premises depreciation	8,000		
	155,000		
NET MARGIN (PROFIT)	25,000		
	180,000		180,000

According to the above example, the profits from the cereal bank amount to 25,000 UC for the year.

The Management Committee must decide what it will do with this profit. This amount could enable, for example, the Association to subsidize jointly the establishment of a new cereal bank in the neighbouring village.

B. MANAGEMENT OF A RICE MILL

A Rice Mill can be enormously beneficial to the people of a grain producing village. Where many privately owned mills charge high milling fees too high priced for the average villager, it is crucial to have a mill owned by the Association, charging a fair price for its operations.

1. The Management Committee

After the decision is taken to establish a mill, the following steps should follow :

a) A Management Committee should be set up for the Mill. It should be composed, for example, of about six men and six women, in order to get a sizeable number of people to participate in the management.

b) A decision should follow on finding the necessary finances to purchase the mill.

First those interested in the project can participate with 100 units of currency as a membership fee.

The balance finances should be found either by borrowing or by asking for a grant from an International donor agency.

c) The Management Committee should take all the major decisions such as selecting the miller, allocating responsibilities, fixing prices, costing the depreciation.

d) The milling price for a KG or a (fixed) measure is established by the Committee depending on the competitive status of other commercial mills. Avoid pricing at a loss. Hence, first make a Profit and Loss Account on the basis of estimated income and expenditure.

e) The Committee establishes, if it deems it necessary, a set of regulations concerning the management and the control of the mill.

f) The Committee's Secretary maintains a book of Minutes of Meetings recording all the decisions taken by the Committee.

2. Technical and maintenance problems

a) The first problem to be faced is in the selection of the mill. Learn from other groups about their experience with different types of mills and heed their advice.

b) With the people's participation, construct a building for the mill. The doors should be lockable.

c) Stock petrol and oil and some spare parts.

d) The main problem will be to MAINTAIN the mill. Two people should
be in charge of it: the miller, and to handle difficult repairs - a
mechanic.

The profitability of the mill depends on its maintenance. If the mill
cannot be used after 4 or 5 years, the investment will be a loss.
On the other hand, if it lasts 8-10 years it becomes an exception-
ally good investment and the price of milling could be reduced as
the mill does not have to be depreciated after the 5th year.

e) It is advisable to have a maintenance book for the mill in which the
miller and the mechanic enter all the checking-up, repairs and
other operations relating to maintenance. The date of each operation, the amount spent and other observations should also be
recorded in the maintenance book.

3. The staff

It is the duty of the Miller to operate the Mill everyday. He starts
operations at a specific time decided upon by the Committee.

He attends to the mill, keeps stocks of petrol, grease and spare parts,
and also keeps the accounts.

The mechanic attends to repairs, carries out regular checks and keeps
the engine, mill and its equipment well maintained.

The mechanic has time to attend to several mills. Therefore, the
Association apportions the number of hours he works for each group
having a mill and correspondingly pro-rates his payment among the
mill-owning groups.

4. Running of the mill

The quantity of rice or other grain that is milled each day should be
recorded for each month and for the entire year.

If the mill has weighing scales, the grain could be weighed in kg or,
as is often the case, in traditional measures.

BOOK OF QUANTITIES MILLED/MILLING CHARGES

RICE MILL

Date	Number of kg or Measures milled		Unit Rate	Amounts Received	Petrol Used
	Type	Kg			
10.6.88	Rice	1500	0.50	750.0	4 lt
11.6.88	Grain	1120	0.60	672.0	3 lt
12.6.88	Cereal	1100	0.50	550.0	3 lt

The daily milling should be entered in another register.

A monthly record form should also be maintained. The addition of the totals of the 12 monthly forms will facilitate accounting of :

- the number of kilos milled annually as well as monthly
- the amount of money received monthly and annually
- the number of litres of petrol used.

In addition a book of daily, montly and annual receipts should be maintained. An example of such a form is given below:

RECORD OF RECEIPTS
Rice Mill

Month / Days	Jan.	Feb.	Mar.	Apr.	May	June	July	Aug.	Sep.	Oct.	Nov.	Dec.
1												
2												
3												
4												
5												
6												
7			etc.			etc.			etc.			
.												
.												
.												
30												
31												
Totals												

To arrive at the expenses enter all the operations that take place in the following manner:

RECORD OF EXPENSES

Rice Mill

Date	Details	Accounts No.	Amount spent	Remarks
15.6.88	Purchase 200 lt. petrol	4	2,700.00	Colombo
15.6.88	Purchase 20 lt. oil	5	160.00	Colombo
21.6.88	Miller's allowance	6	500.00	
22.6.88	Repayment of loan	7	1,500.00	People's Bank Puttalam
25.6.88	Spare parts	8	3,000.00	Martin's Garage

5. Available cash

The mill charges money from each person milling grain.

Each evening the miller hands over the cash to the Treasurer together with daily receipt form duly completed. The Treasurer signs and keeps the money in a safe place.

When the collections in the cash box exceed a stipulated limit (5,000 units of currency) the money should be deposited in the Bank.

The signatures of both the Executive Secretary of the Association and the Treasurer are necessary to withdraw large amounts of money from the bank for meeting expenses.

Note: The Treasurer should never keep more than 5,000 UC in the box because of the risk of theft. Furthermore, this money will be idle and will not earn any interest if not invested.

6. How to work out the depreciation of the mill

If the mill costs 150,000 UC and the building 25,000 UC the total investment is 175,000 UC. It is estimated that provided the mill is satisfactorily maintained it can run for 8 years, but the Committee decides to depreciate it in 5 years. Therefore each year the Committee should set aside:

$$\frac{175,000}{5} = 35,000 \text{ UC}$$

to depreciate the mill.

7. Annual profit or loss

To determine whether the group is making a loss or a profit, take into account all items of income and expenditure:

Examples of Income:
- membership fees of the group running the project
- the amount collected as rice milling charges
  (as per monthly income form)
- bank deposit interest
- sale of other milled cereals

Examples of Expenses
- petrol, oil and grease
- transport and travelling costs
- allowance paid to the miller and the mechanic
- all other running costs
- annual depreciation

The Annual Profit and Loss Account of the mill can be worked out. An example follows :

Example

RICE MILL

Profit and Loss Account
for year ending 31st December 1988

EXPENSES:			INCOME:		
Petrol		40,000	Milling for the year (for 12 months)		180,000
Maintenance items,		20,000	Bank Interest		12,000
Transport		10,000	Miscellaneous sales of other grains		60,000
Allowances					
Miller	36,000				
Mechanic	24,000	60,000			
Purchase of grain		48,000			
Annual depreciation		35,000			
		213,000			
PROFIT		39,000			
		252,000			252,000
		========			=======

Thus, after depreciation, the net profit of the group running the mill is 39,000 UC.

It is the responsibility of the Committee to decide what it will do with the profits. It can either decide to increase the depreciation charged or to keep the profit as a reserve in the Bank. It can also decide to help another village to buy its own mill by granting them a loan or donation.

PART FIVE

ACCOUNTING SYSTEMS

A. KEEPING ACCOUNTS

An Accounting System provides information about income and expenditure, cash in hand and funds in the bank, the distribution of the expenses and income under various categories, and the outcome of operations. It will also indicate whether the enterprise shows a profit or a loss at a given date for a specified duration of time.

For example, when purchasing stamps (25 UC) to airmail a financial request to a donor, a double entry should be made in the accounts:

- expenditure taken out of the cash box
- increase of the general expenses (stamps).

> Thus, an expenditure "OUT" from one side and "IN" from the other, is commonly known as <u>double entry accounting</u>

The "OUT" and "IN" should be recorded in a "T" Account according to this simple formula:

Or if membership fees are received from a member (100 UC):

Therefore, no two entries in the double-entry system can be on the same side of the account. There must always be a balance between <u>debit</u> and <u>credit</u>.

B. ACCOUNTING PROCEDURES

The recording of accounting entries follows an operational logic which could be presented thus:

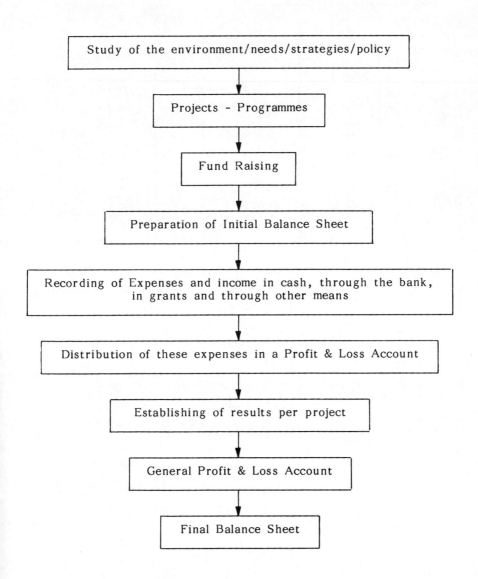

The general outline of an accounting system for a development association could also be presented as follows overleaf.

C. ACCOUNTING SYSTEM OF A DEVELOPMENT ASSOCIATION

Study of the Environment --- Needs --- Strategies --- Policy --- Programme --- Project

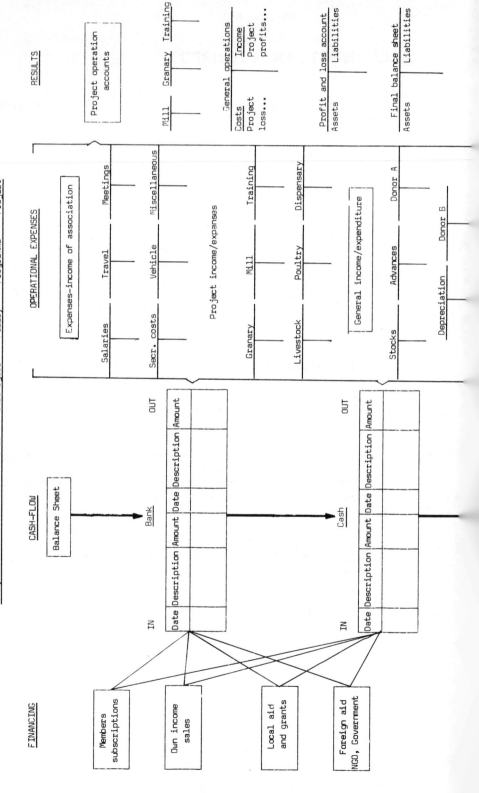

D. ACCOUNTING METHODS

There are various methods of keeping accounts, moving from the very simple to the more complex.

The simplest system is to enter the expenses on the right hand page and the income on the left hand page. An example follows :

ACCOUNTS BOOK
(on two pages)

| Left page | | | | Right page | |

EXPENSES   INCOME

Date	Description	Amount	Date	Description	Amount

Accounts can be also kept on one page but in three columns: income, expenses and balance.

Each accounting document should be numbered and when making the corresponding entry that same number should be recorded in the corresponding column for easy re-tracing.

ACCOUNTS BOOK
(on one page)

Date	Description	The accounts Doc. No.	Counterpart Account No. (double entry)	Debit IN	Credit OUT	Balance

● Accounts can also be kept on printed forms which can be purchased from stationers.

When using such forms, it is only necessary to fill in the year, date, description, expenditure, income and balance. The voucher number and possibly the corresponding account number can also be filled in if the accounts are in double entry.

● There is also an "American" accounts system which, at first sight, seems simple but is not very practical if many details are needed.

On the other hand, if all the accounts can be summarised in 10 accounts, the American system is useful.

EXAMPLE OF THE AMERICAN ACCOUNTING SYSTEM

DATE	DESCRIPTION	TOTAL	CASH		BANK		PROJECTS							
							MILL		TRAINING		CEREAL BANK		CULTIVATION	
			DEBIT	CREDIT	DEBIT	CREDIT	DEBIT	CREDIT	DEBIT	CREDIT	DEBIT	CREDIT	DEBIT	CREDIT
1.1.88	INITIAL BALANCE	120,000	36,000		84,000									
5.1.88	PURCHASE SEEDS	3,000		3,000									3,000	
6.1.88	PURCHASE TOOLS/Mill	600		600			600							
1.2.88	PURCHASE WATER PUMP	13,000				13,000							13,000	
8.2.88	MAINTENANCE OF MILL	5,000		5,000			5,000							
9.2.88	PAYMENT FOR FERTILIZER	500		500									500	
ETC.		etc.				etc.								

The Debit/Credit entries should be equal for each operation.

- Finally, if the Association has a large budget, it is advisable to use the computer of a specialised organisation in the city. In this case three journals should be maintained:

    - cash journal
    - bank journal
    - miscellaneous operations journal

It is necessary to write alongside each registered line, the account number in the system to be debited or credited. The computer will do the distribution and the calculations.
(The advantage in using a computer is that there can be no mistakes in the calculations).

The accounting system that is selected should be well mastered by those in charge of the Association and it should be simple and clear enough for the members to understand, and for them to use for maintaining the basic accounting documents such as the cash book of the mill, the group's bank account and the stock forms.

E. CLASSIFICATION OF ACCOUNTS

Classifiying and re-grouping all the accounting data helps to prepare the financial statements. Every account of the Association is identified by a code in order to:

    - simplify the classification and quickly identify each account
    - specify the class and series to which different accounts belong.

The system recommended below is suitable for use in the management of a local association.

The accounting plan follows a DECIMAL CLASSIFICATION. Here is a simple example:

The first figure, 5 , represents the cagegory of all financial accounts.

The figures 5.1, 5.2, 5.3 etc. represent the sub-classification of these accounts. Thus we have:

5 - Financial Accounts

    5.1   Cash
    5.2   Bank
          5.2.1 Savings and Credit Account
          5.2.2 Current Account
          5.2.3 Fixed Deposits

The decimal classification can be adapted according to the requirements and the organisation of the activities by adding on decimals when sub-classifications are needed.

When a central account is sub-divided according to member groups, the classifiction can be made in the following way:

5 - Financial Accounts

    5.1   Central cash account

          5.1.1 Rice Mill cash
          5.1.2 Cereal Bank cash

               5.1.2.1 Cereal Bank A cash

               5.1.2.2 Cereal Bank B cash

               5.1.2.3 Cereal Bank C cash

    5.2   Bank Account

    etc.

Note that only nine figures are available for classification.

CLASSIFICATION OF ACCOUNTS FOR A DEVELOPMENT ASSOCIATION

Class 1   CAPITAL AT MEDIUM OR LONG-TERM

    1.1   Capital
    1.2   Reserve funds
    1.3   Temporary assets
    1.4   Temporary liabilities

Class 2   FIXED ASSETS

    2.1   Land
    2.2   Buildings
    2.3   Equipment
    2.4   Vehicles
    2.5   Machines
    2.6   Livestock

Class 3   STOCK

    3.1   Raw material and supplies
    3.2   Stock of goods and finished products

Class 4   THIRD PARTY ACCOUNTS

    4.1   Suppliers
    4.2   Debtors
    4.3   Advances to staff

Class 5   FINANCIAL ACCOUNTS

    5.1   Cash
    5.2   Bank
    5.3   Deposits

Class 6   ASSOCIATION'S EXPENSES

    6.1   Staff salaries and charges
    6.2   Rental
    6.3   Maintenance and repairs
    6.4   Secretariat expenses
    6.5   Communication costs
    6.6   Meeting expenses
    6.7   Taxes
    6.8   Interests

Class 7   INCOME

   7.1   Membership fees
   7.2   Grants
   7.3   Sales revenue

      7.3.1   Granary
      7.3.2   Mills
      7.3.3   Breeding/Livestock
      etc

   7.4   Fees and services

Class 8   PROJECT EXPENSES

   8.1   Wells
   8.2   Roads
   8.3   Rice mills
   8.4   Collective fields
   8.5   Cereal Bank
   8.6   Livestock
   etc.

Class 9   SOCIAL EXPENSES

   9.1   Training Programmes
   9.2   Literacy Programme
   9.3   Health and Social help
   9.4   Cultural Activities
   etc.

N.B.: Example of detailed classification

8.4. Collective fields
  8.4.1. Collective field Wilpotha
    8.4.1.1. Seeds
    8.4.1.2. Fertilizer
    8.4.1.3. Transports
    8.4.1.4. Materials
    8.4.1.5. ...
    8.4.1.6. ...
    8.4.1.9. Amortization

  8.4.2. Collective field Anapura
    8.4.2.1. Seeds
    8.4.2.2. Fertilizer
    8.4.2.3. etc...

Once the mechanics of the system are understood, the Association can design its own accounting classification by adapting the titles of each category or sub-category of expenses or income.

PART SIX

PROFIT AND LOSS
ACCOUNTS

A. PRELIMINARY BALANCE SHEET

When an Association organises its accounting system, it has to first prepare the balance sheet, using the inventory of available assets and liabilities. The following are examples of available assets and liabilities:

- fixed assets: land, buildings, furniture, machinery, equipment, material or other fixed assets
- liquid assets: cash in hand or in bank
- amounts which are due to the association or which the association owes to outsiders (debtors and creditors).

Once everything has been noted and the corresponding values estimated, list all the assets on one side and liabilities on the other side.

An example of a preliminary budget estimate is given below:

PRELIMINARY BALANCE SHEET
OF A RURAL DEVELOPMENT ASSOCIATION
as at 30.6.87

in UC

ASSETS			LIABILITIES	
Description		Amount	Description	Amount
Assets				
motor pump	13,000			
machinery	5,000			
equipment	10,000	28,000		
			Members' contribution	18,000
			Grant for equipment	15,000
Bank		10,000	Loan	6,000
Cash		1,000		
		39,000		39,000

A typical Balance Sheet of a medium sized Development Association will
be presented in the following way:

A STANDARD BALANCE SHEET

OF AN ASSOCIATION

in UC

ASSETS		LIABILITIES	
Fixed assets: land, buildings, equipment, machinery, livestock	____	Capital at medium and long-term Profit and Loss account	____ ____
Stocks: raw materials, finished goods	____	Reserves	____
Debtors and advances to staff	____		
Cash in hand Cash in bank	____ ____		
Temporary assets	____	Temporary liabilities	____
TOTAL	xxxxx equal	TOTAL	xxxxx equal

B. THE DISTRIBUTION OF INCOME AND EXPENDI-TURE

1. The distribution system

Almost all items of income and expenditure of an association are re-corded in its CASH and BANK books. It is important to maintain these books accurately as all the financial information required is recorded in them.

Thus, it is from the CASH and BANK accounts maintained by the villagers and monitored by the Treasurer of the Association that the AMOUNT SPENT ON EACH PROJECT is extracted.

CASH BOOK NO. 5100   in UC

DATE	DESCRIPTION	ACCOUNT NO.	DOCUMENT NO.	DEBIT	CREDIT	BALAN
1.1.88	BALANCE SHEET		1	1000		100
3.1.88	PURCHASE SHOVEL	8219	2C		30	97
4.1.88	TRAVEL TO TOWN – TRAINING	3500	3C		15	95
6.1.88	SALE OF VEGETABLES	8219	4C	150		11
8.1.88	TRANSFER TO BANK	5200	5C		750	3

BANK BOOK NO. 5200   in L

DATE	DESCRIPTION	ACCOUNT NO.	DOCUMENT NO.	DEBIT	CREDIT	BALA
1.1.88	BALANCE SHEET		1	15000		156
5.1.88	PURCHASE FERTILIZER	8219	2		500	14
8.1.88	PURCHASE WATERING CANS	8219	3		700	13
8.1.88	DEPOSIT-CASH	5100	5C	750		14

COLLECTIVE FIELDS NO. 8219

in UC

EXPENSES				INCOME		
3.1.88	2C	SHOVELS	30	6.1.88	4C	SALE OF VEGETABLES 150
5.1.88	2	FERTILIZER	500			
8.1.88	3	WATERING CANS	700			

TRAINING NO. 3500

EXPENSES				INCOME
4.1.88	3C	TRAVEL TO TOWN	15	

All cash and bank entries are thus allocated to the corresponding accounts. Double entries; otherwise the accounts will not balance.

1. Individual cash and bank books relating to mills, collective fields, etc. should be kept in the village.

2. In the Secretariat of the Association, the Treasurer should open a form either for each corresponding account or else for each village group.

2. The accounts relating to a programme

Programme accounts lie within the framework of the general accounting plan of the association.

Each programme will be entered in classes 8 or 9 according to its nature (refer to page 74 - Classification of Accounts for a Development Association), which means that the first figure of the account number of the programme will start with 8 if it is a productive programme and with 9 if it is a social programme.

Take for example the accounts of a rice mill operated by an Association.

EXPENSES	INCOME
- petrol, oil and grease	- milling charges
- maintenance supplies: millstones, filters	- other sales - bank account interest
- transport costs	
- allowances to the miller and to the mechanic	
- annual depreciation	
- other expenses	

Allocate numbers to each of the accounts operated.

● For the income of the Rice Mill the classification will be 7.3.2.0

1st figure: 7 - corresponding to income
2nd figure: 3 - corresponding to the particular village
3rd figure: 2 for instance, corresponding to the mill
4th figure: 0 = total income

The particular mill will therefore be registered under No. 7320

● For the Expenses of this particular Rice Mill the classification will be as follows:

> 1st figure: 8 corresponds to a productive project which is the production of a service
> 2nd figure: 3 will correspond to the particular village = 3
> 3rd figure: 2 corresponds to expenses of all the mills and 832 to the particular mill
> 4th figure: this will be the classification of detailed expenses
> 8.3.2.1 : petrol, oil
> 8.3.2.2 : maintenance supplies: e.g. millstone
> 8.3.2.3 : transport
> 8.3.2.4 : miller's and mechanic's allowances
> 8.3.2.5 : miscellaneous
> 8.3.2.6 : depreciation

Some other examples will help explain this classification system further:

The fuel expenses of mills located in villages other than the above will have the following numbers 8421, 8521, 8621. It is the 2nd figure that changes according to the village or zone.

Or again, the transport expenses of the above-mentioned granary = 8363, the granary of another village will be = 8463.

The Accountant should decide on the structure of expenses and in the case of all projects he should maintain the same numbering for "Maintenance Costs", "Production Costs", "Salaries", etc. Such a numbering system will simplify comparisons and enable regrouping of expenses/-income. For example, it will enable the regrouping of the salaries paid to the entire staff of the association.

3. Accounts of the Coordinating Secretariat

In the general accounting, the expenses relating to the Association's general administration, i.e. the expenses of the Secretariat or of Coordination should also be recorded.

The expense categories of a Secretariat responsible for the animation or coordination of a development association are, for example:

- salaries and social expenses
- travel and contributions to groups
- communication costs; stamps, telephone, despatch, etc.
- secretariat costs: office supplies, rental, equipment, cost of meetings, etc.
- other general expenses
- depreciation of equipment.

The following example can be adapted to meet the requirements of an internal accounting plan:

Example:

1st figure - 6 = SECRETARIAT EXPENSES

2nd figure - 0 = This column is not really necessary but since all the numbers of the accounting plan have four digits, use 0 everywhere.

3rd figure - 1 = salaries and social expenses
- 2 = travel/transport
- 3 = communications costs
- 4 = secretariat costs
- 5 = meeting expenses
- 6 = general expenses
- 7 = equipment

4th figure - sub-categories of 3rd figure

e.g. communication costs

- 6.0.3.1 = stamps
- 6.0.3.2 = telephone
- 6.0.3.3 = despatch

The income of the coordinating secretariat is generally small and it is well below the expenses. The Executive Committee is always confronted with problems of financing these expenses. Previous chapters have dealt with this problem.

Apportion a part of the secretariat expenses to projects, making use of the percentage key-systems. It can be estimated, for example, that 20% of the coordinator/animator's allowance during the past year should be apportioned to "cultivation activities". In this case add a "Common Management Costs" item equal to 20% of the animator's salary to the "general expenses" of the "Cultivation Account".

It is possible to do the same with another animator whose expenses have been accounted in the general secretariat account. The same applies for costs of meetings, communications, etc.

Hence it is advisable to distribute the costs of administration/coordination/animation in such a way as to increase the costs of projects. It is easier to have projects funded than to find a donor who will be willing to support the administrative/coordination overheads of an association.

C. DEPRECIATION

Depreciation is calculated by dividing the value of some equipment or investments by the number of years estimated to be its lifetime. If a van purchased at a cost of 150,000 UC, is carefully maintained and is expected to last at least 5 years, then the annual depreciation of this van will be:

$$150,000 \quad - \quad 5 \quad = 30,000 \text{ UC}$$

The following depreciation table can be drawn for this van:

in UC

YEAR	PURCHASE	ANNUAL DEPRECIATION	CUMULATED DEPRECIATION	BALANCE
0-1988	150,000			
1-1989		30,000	30,000	120,000
2-1990		30,000	60,000	90,000
3-1991		30,000	90,000	60,000
4-1992		30,000	120,000	30,000
5-1993		30,000	150,000	-

Each year a sum of 30,000 UC should be entered in the expenses account and the Profit and Loss account as depreciation on the van. If annual depreciation is not entered, the Profit and Loss account will not show the correct picture and the van may be eventually sold at a loss.

Depreciation techniques

Do not confuse depreciation and reserves/provisions for replacement.

Depreciation is purely an accounting notion; e.g. if a machine with a lifespan of 3 years is purchased for 30,000 UC, appropriate 10,000 UC to the various activities which it will serve. The accounting depreciation is therefore an expense which is spread over a number of years and of which the annual cost will be entered as costs to the relevant activity.

The reserve for replacement of equipment and fixed assets is in effect a provision that is made in the bank account so that the necessary funds are available to replace equipment and machines without having to seek more foreign aid.

The easiest technique is to open a bank account called "Reserve for Replacement of Assets". Each year, at the closing of accounts, deposit "cash" in this account corresponding to the amount provided for during the year for replacing fixed assets.

No withdrawal is possible from this account other than when the General Assembly or the Committee which has been mandated for this, decides to replace the assets (machines or equipment) in question.

Making entries

There are two ways of recording annual depreciation in the accounts:

a) First Method: open an account and enter the purchase value as a debit and the depreciation as a credit. Thus the accounting value of the equipment is obtained by calculating the balance in the account, at any particular date.

	1988		
DEBIT	VAN		CREDIT
	UC		UC
Purchase	150,000	Depreciation	30,000
Balance at the end of the year	120,000		

b) Second Method: open two accounts:

- first, an account to register as a debit the purchase price and other investments of all fixed assets which are being depreciated

- and a second account in which the annual depreciation of the fixed assets and other investments is recorded as a credit.

This method has the advantage of showing in the balance sheet the purchase price of your machinery, equipment or other investments which are being depreciated.

The provision for replacement which will be entered in the "investment" bank account will appear as follows:

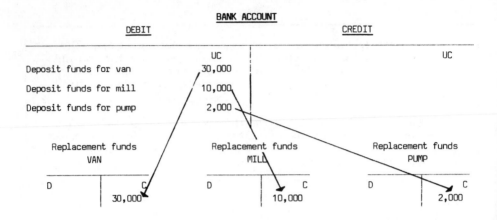

The <u>final balance sheet</u> will show the total amount of money in debit in the bank for the replacement of equipment and machinery; and the same amount in credit in the total replacement account.

When a sum of 150,000 UC is withdrawn from the bank to purchase a new van the following entry should be made:

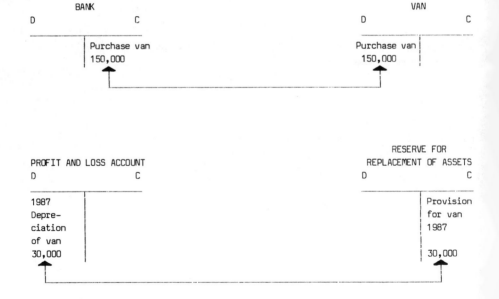

D. COST CALCULATIONS

The calculation of production costs will disclose whether the projects (mill, collective fields, etc.) earn money or not. However, consider that profits are not only financial but also social and that a project can quite well lose financially on condition that an important "social benefit" is gained. However, the Association should have the means to cover such financial loss as the final aim of a project is social as well as economic.

<u>Use of Accounts</u>

Double entry accounts will show the exact annual cost of the activity undertaken. For example at the end of the year, if the expenses on account of mill A total 100,000 UC, the total cost divided by the number of kilos milled will give the cost per kilo milled.

$$100,00 \text{ UC} \div 50,000 \text{ kg} = 2 \text{ UC per kg}$$

Any annual income over 100,000 UC per year is a profit.

Any annual income less than 100,000 UC is a loss.

> Use the accounts to <u>manage</u> the projects and programmes as well as for decision making; not only to justify the foreign aid received.

1. Example of a cost calculation of handicrafts manufactured by a group (weaving)

> Purchase price of raw material (e.g. wool/cotton)
> + Transport (travel)
> + Miscellaneous purchasing costs
> _____
>
> = FINAL PURCHASE PRICE
>
> + Added value, i.e. the value of work done to produce
>   the item or the service (number of days X daily wage)
> + Complementary products added to the initial product or
>   maintenance of product (buttons, zips, dyes, wire, etc.)
> + Miscellaneous manufacturing costs
> + Depreciation
> + Apportioned management costs of Association
> _____
>
> = TOTAL COST PRICE
>
> + Profit margin
> _____
>
> = SELLING PRICE (divided by the number of pieces if necessary)
> ==================

2. Apportioning fixed charges and general expenses per project

This is a delicate operation which must be carried out if the exact cost of any single activity is to be known.

In fact, if no one finances the general expenses, salaries or the technical framework, or if a funding organisation has so far granted aid and stops, a solution has to be found without jeopardising the activity and thereby the Association.

Allocate to each project the expenses corresponding to the quantity of products or raw material which they would have used.

Pick out what is easy, e.g. number of bags, the quantity of petrol, which will be recorded in the stock account. It should be immediately debited and carried over to the expenditure of the project in question.

However, it is not possible in some expense categories to know exactly the amounts which should be appropriated to the various activities concerned.

So work out a cost scale calculated on a percentage basis.

For example, take the cost of the vehicle used by the mechanic to maintain the mills, pumps and other equipment belonging to the association.

Make a list of projects which have used the services of the mechanic, who has travelled from project to project using the vehicle, and apportion the costs among the various projects.

For presenting the general expenses such as secretariat costs, communication costs, animators'/coordinators' salaries, etc., the best method is to calculate these costs as a percentage of total activity costs. Thus if the annual total costs of an association are 1,200,000 UC and the annual general expenses are 120,000 UC, then the general expenses compared to the total project activities will be :

$$\frac{120,000}{1,200,000} = \frac{1}{10} \text{ or } 0.1 \text{ or } 10\%$$

Therefore add 10% of the general expenses to the direct costs to enable the calculation of the total cost of a specific project.

E. THE PROFIT AND LOSS ACCOUNT

To know if a profit or loss has been made, prepare a Profit and Loss Account per category of activities and a general Profit and Loss account for the Association.

1. Take the example of the cereal bank again (page 59)

THE PROFIT AND LOSS ACCOUNT
OF THE CEREAL BANK
FOR THE 12 MONTHS ENDING 31 MARCH 1988

EXPENSES/COSTS	UC	INCOME	UC
Purchase of cereals	125,000	Sales of cereals	175,000
Transport	6,000	Bank interest	5,000
Loss of grain	2,000		
Storekeeper's allowance	10,000		
Maintenance	4,000		
Premises depreciation	8,000		
	155,000		
NET MARGIN (PROFIT)	25,000		
	180,000		180,000

The cereal was bought for 125,000 UC and sold for 175,000 at a profit of 50,000 UC. It is this gross profit which must cover the expenses and generate a net profit. The expenses of the cereal bank during the year comprise the cost of transport, loss of grain, miller's allowances, maintenance, and depreciation of premises: a total of 30,000 UC.

Taking into account the interest from the bank, the profit will increase by 5,000 UC - i.e. to 25,000 UC.

The Association should take into account the initial and final stocks, purchases and sales during the year, cost of inputs, and operational costs, distribution of general expenses per project, and the depreciation of buildings, machines, material and furniture, before concluding whether the year's activities should be seen as profitable or not.

The Committee of the Association should discuss how to use the profit.

The profits can be used to:

- cover the deficits of other projects;
- depreciate all the buildings, machinery, equipment and furniture;
- make necessary reserves as provisions for the future;
- distribute the money equitably (in return for their efforts) among all those who participated in achieving the profit, proportionate to the efforts that each one has made.

2. The General Profit and Loss Account

At the end of the year (it should correspond with the end of the agricultural year) the general profit and loss account of the Association should be estimated, for example:

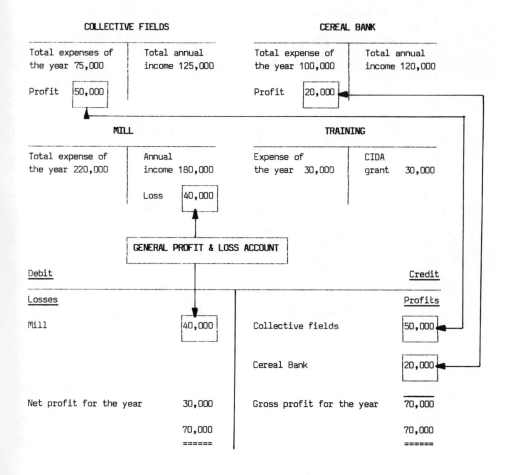

Thus, after closing of all accounts and summing up the results of each activity in the general profit and loss account, the above example will show that in 1988 the profit was 30,000 UC.

By analysing the above figures it can be seen that it is the collective fields and the cereal bank which are keeping the Association alive. The costs of running the mill are high and it has registered an annual loss of 40,000 UC. A decision should be taken at this stage about the future of the mill.

F. THE FINAL BALANCE SHEET

The Balance Sheet is the table of assets and liabilities of an Association, project or enterprise.

On the left, is a list of the Assets.

On the right, is a list of the Liabilities.

The difference between the Assets and the Liabilities will give the real value of the CAPITAL.

It will be advisable to make out the final balance sheet keeping in mind the modifications made by the allocation of expenses and income of the year.

The Balance Sheet Assets are:

- the amounts available in the bank and in the cash box

- the value of assets: land, buildings, vehicles, machines, furniture, equipment

- the value of stock (raw material or end product) at the end of the accounting year

- loans or other advance payments given to members and bills upaid by customers.

The Balance Sheet Liabilities are:

- the group's debts: unpaid bills of the previous year

- unutilized grants

- reserves for replacement of material

- own Capital which will vary according to the results of each year (whether profit or loss)

EXAMPLE
FINAL BALANCE SHEET
OF A DEVELOPMENT ASSOCIATION

as at 31.12.1988

ASSETS			LIABILITIES in UC	
FIXED ASSETS (net value)			CAPITAL	200,000
* land	50,000			
* buildings	100,000		RESERVES	200,000
* vehicles	60,000			
* machinery	75,000			
* furniture	15,000			
		300,000	DEBTS/LOANS	200,000
STOCKS		210,000		
			GRANTS NOT	
LOANS, ADVANCES		15,000	YET USED	110,000
CASH IN BANK	150,000			
CASH	35,000			
	185,000			
TOTAL		710,000		710,000
		=======		=======

Here is another example:

BALANCE SHEET

UC

Assets (Wealth)   Liabilities (Debts)

Fixed assets	-----	Capital	-----
Land		General reserves	-----
Buildings			
Machines		Balance profit & loss account	-----
Vehicles			
Equipment		Provision for replacement	-----
Livestock			
		Creditors	-----
Stocks	-----	Temporary liabilities	-----
Cereal Bank		Grants received in advance	_____
Petrol			
Collective fields			
Deposits/Investments	_____		
Deposit in bank			
Other investments			
Cash-flow/advances	-----		
Bank current accounts			
Debtors			
Invoices due			
Advances to staff			
Temporary assets	-----		
TOTAL	XXXXX	TOTAL	XXXXX

The final balance sheet should be prepared by a competent accountant. It is advisable to seek the assistance of a qualified accountant who could also help in organising the accounting system of the Association.

The accounts will then not only be a tool for management of the funds received but also a tool for forecasting.

PART SEVEN

FINANCIAL CONTROLS AND JUSTIFICATION OF EXPENSES RELATED TO GRANTS

A. JUSTIFICATION OF FUNDS RECEIVED FROM DONORS

1. The donor's account

It is sometimes a condition of a grant that the Association maintains a specific donor's account, which should be maintained in the same manner as the cash book.

For example:

CIDA (Canada)

Contract : 100,000 UC for 2 years

DATE	DETAILS	ACCOUNTS DOCUMENT NO.		IN	OUT	BALANCE
1985						
8 Jan.	Advance received from CIDA	1	Bank	100,000		100,000
15 Jan.	Cereal bank Purchase of cement	2	Bank		25,000	75,00
18 Jan.	Cereal bank Purchase of stones	8	Bank		5,000	70,000
18 Jan.	Cereal bank Purchase of sand	9	Cash		3,000	67,000

Record in the "IN" column the grants received via the bank and record all the expenses in the "OUT" column of which the accounts documents have to be sent to the donor.

Thus, at the end of the year (or whenever needed) it will be easy for the treasurer to establish the list of all the expenses incurred from the donor's grant and to furnish all the accounts documents (numbered) corresponding to the expenses. If the accounts are well kept, it is best to photocopy them and send a copy to the donor.

2. Documentary proof to be submitted to the donor

It is not only a contractual obligation, but also a moral obligation to provide information on how the money received as a subsidy/grant was utilized.

Here is an example of the items that should be included in a dossier that is to be prepared for submitting to a donor after utilizing a grant.

This dossier should consist of:

a) A covering letter submitting the documents and thanking the donor for the aid received (1 page)

b) A report covering the activities and results achieved after having used the grant (3 to 10 pages, depending on the case)

c) Financial documents in summary (1 page)

d) A detailed financial statement of expenses concerning the aid received (1 to 5 pages depending on the case)

e) Annexes such as:

(i)   the general accounts of the Association
(ii)   the general report of the Association
(iii) photos, drawings, plans

Refer to Annexe VIII.1: Specimen of accounts to be submitted to the donor
(a) Covering letter
(b) Detailed financial statement
(c) Summary of the financial assistance received

3. Contents of a project report relating to the grant received

This document should describe in detail the facts which correspond to the activities carried out with the grant:

It should also comprise:

- the name and address of the Association
- the date of the report
- the exact nature of the project
- the date of commencement of the activities
- the signature of the Chairman/Secretary of the Association

Contents of the report

This report, submitted as an annexe, should give the following:

(a) the objectives of the project

(b) achievements during the period

(c) problems and difficulties encountered and the solutions

(d) how the project responded to the local requirements and its impact on the people, the Association members and the villagers

(e) Annexes - useful documents to back up the report (but only concerning the project funded by the donor and nothing else).

B. FINANCIAL CONTROL / AUDITING

1. Auditing should start by checking the currency in the cash-box and the money in the bank.

   a) Reconcile the cash-position according to the cash-book (the day's indicated balance) and the total money available in the cash box on that day.

   Refer to Annexe VII.2 Cash control form

   b) For the Bank audit compare the balance according to the Association's accounts with the bank statement and make the necessary reconciliation entries (interest, charges, or cheques which are not yet declared).

   Refer to Annexe VII.3 Bank Reconciliation form

   c) The most important scrutiny will be of the expense vouchers submitted by the Executive Committee members or any other persons.

   When an advance is given to a person, he should sign a receipt for the amount of this advance. The Treasurer should keep this receipt until he produces a written statement of all the expenses incurred by him. He should support all these expenses with a bill or a receipt and he should write on a separate page all the expenses for which he could not get a receipt (taxi, telephone, stamps, etc.).

Annexe VII.4 gives an example of a statement of expenses any person of an association should submit when money is advanced for a specific purpose.

d) The <u>statement of equipment</u> according to the inventory should be next audited *(refer to Annexe VII.5: Specimen inventory of goods)* and necessary decisions should be taken.

2. The Secretary-General, the Director or the Financial Manager should exercise budgetary control over the general accounting of the association.

This control is exercised through the comparison of the expenses estimated in the budget with the actual expenses incurred during the year. This control should be exercised monthly or quarterly and should be followed up with appropriate decision making.

3. Auditing of Accounts

In a local development association, the consitution makes provision for the accounts to be audited regularly. There are two ways of auditing; either through internal auditors selected from among the members or by obtaining the services of an external Auditor.

Internal auditors should be qualified persons; very often it is difficult to find such persons in a local development association.

After auditing the accounts, the auditor (internal or external) should submit an audit report which certifies to the donors and other concerned outsiders that correct and proper accounts have been maintained by the Association.

Using the services of a qualified Chartered Accountant is recommended.

Refer to: Annexe VII.6   A List of Necessary Controls to be effected within an Association.

ANNEXES

SPECIMENS & EXAMPLES

Several specimens and examples of documents used by Development Associations designed by their Management through past experience are given in the following annexes. Several case studies have also been quoted.

Make use of these to create instruments of management most suited to your requirements.

Explanation of the classification:

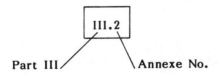

Part III   Annexe No.

ANNEXES PART I

FINANCING AN ASSOCIATION AND ITS ACTIVITIES

LIST OF PERSONS AND ORGANISATIONS
WHO COULD SUPPORT THE ASSOCIATION

	NAME ADDRESS	TELEPHONE No.	REMARKS
TECHNICAL SUPPORT			
– Agriculture	Mr. S. Fernand, Director Ministry of Agriculture	23571	foods crops
– Roads, wells	Mr. S Tudor, Engineer	54231	well specialist
– Sanitary installations	Mrs. Niembe		
– Appropriate technology	...		
– Management	etc.		
GOVERNMENT SERVICES			
– Ministry	etc.		
– Local authority			
– Extension service			
FINANCES			
– Government			
– Embassies			
– Internationl Financing Organisations			

CRITERIA FOR SELECTION OF PROJECTS

(An example taken from the guidelines followed by a donor agency)

The following criteria are applied in selecting from the numerous project proposals submitted by third world development associations, those that should be supported with a grant. The criteria, however, are not applied rigidly. The originality, form and motivation under-lying the proposal are very important when it is considered for sup-port by a donor agency.

1. PROJECT REQUIREMENTS

1.1 The project should respond to the requirements of the ben-eficiaries.

1.2 Priority is given to integrated and localised development pro-jects; viz: projects having a functional integration of economic, social and cultural components in addition to having internal economic viability.

1.3 The project should comprise three major objectives:

a) Training/animation - aimed at obtaining the maximum par-ticipation of those concerned and with a view to preparing them for taking over of the project.

b) A quantitative and qualitative raising of the standard of living of the beneficiaries.

c) Having a multiplier effect in the area or the region.

1.4 The project should aim at attaining in the shortest possible time the maximum degree of sustainability of the economic activities envisaged by the project; its viability is not ne-cessarily tied up to its profitability.

1.5 As a general rule, the project should fall within the objectives of the national development plan.

1.6 The project manager need not invariably be a government official.

1.7 Beneficiary participation is necessary.

1.8 The financial participation of the (donor) organisation should not exceed, for example, 60,000 UC per year per project. A small grant not exceeding 5,000 UC may be given for a preli-minary study to determine the possibility of such a project.

2. CONCEPT

2.1 The project should be compatible with the needs and the abilities of the people.

2.2 The project should aim at reducing the level of economic and managerial dependence of the beneficiaries on external structures.

2.3 Priority will be given to projects using an appropriate technology and local resources with material and personnel.

2.4 If the project requires the services of expatriates, it should be planned to replace them with local nationals as soon as possible.

3. PROJECT AIMS

3.1 A project should be a vehicle through which people in the donor country are made aware of Third World problems.

3.2 The donor's information or educational committee would be in charge of this task.

The following selection criteria will help to serve as an indication of what projects are considered to be worth financing:

1. REQUIREMENTS

1.1 Requirements expressed by the beneficiaries.
1.2 Integrated and localised development.
1.3 Dual objectives:
   a) training-animation
   b) raising the standard of living
1.4 Rapid autonomy
1.5 Integration into the national development plan
1.6 Management with or without official participation.
1.7 Beneficiary participation
1.8 Grants

2. CONCEPT

2.1 Conformity with the local needs and abilities of beneficiaries.
2.2 An effort to end the system of economic domination and political repression of the masses.
2.3 Use of appropriate technology and local resources.
2.4 Rapid replacement of foreign expatriates by local nationals.

SUMMARY OF A REQUEST FOR FINANCIAL AID
FOR A LOCAL PROJECT

NAME OF PROJECT:

> DRILLING OF SIX WELLS
> TO OBTAIN WATER SUITABLE FOR
> HUMAN CONSUMPTION

ORGANISATION IN CHARGE: Vinivida NGO Coalition
Angunawila, Mundel
Sri Lanka

COST OF PROJECT: 60,000 UC

REQUESTED FROM CIDA: 45,000 UC

LOCAL PARTICIPATION: 15,000 UC

EXCHANGE RATE APPLIED: 1 US$ = 30 UC

BANK ACCOUNT: No. 123456789
at Peoples's Bank, Chilaw

Angunawila, 30 June 1988

SUMMARY OF PROJECT REPORT TO BE FORWARDED WHEN REQUESTING FINANCIAL AID

	No. of Pages
0. Contents	1
1. Introduction of the NGO: history & background Activity report Achievements	1
2. Title of Project and description	1
3. Objectives of Project Global Specific (study of environment, answer to the needs of the people)	2
4. Organisation of the Project Technical Administration Training/Participation	3
5. Means/Resources Human - available/to be obtained Material (list) Financial	2
6. Plan of Implementation - Timetable	1
7. Costs Investments (list prices/depreciation) Operations	2
8. Profitability Profit and Loss account Economic justification	1
9. Financial Plan Contribution of NGO Local Financing Products/Sales Foreign contributions needed	1
10. Funding requested from Donor Agency	1
TOTAL	16
+ Annexes	

SPECIMEN AGREEMENT BETWEEN AN EMBASSY AND A LOCAL ASSOCIATION
to obtain funds for a small rural development project

FUNDING AGREEMENT

Between   The Embassy of ....

and   Name of organisation ....

           Person in charge ....

           Address ....

           Telephone ....

This agreement is in connection with the funding by the Embassy of .... for the following project:

1. Title of project ....

2. Exact location ....

3. Amount funded by the Embassy ....

   Date ....

4. Time schedule:

   a) the project will commence on ....
   b) it will be completed on ....

5. Elements of monitoring

   The person in charge agrees to submit to the Embassy:
   a) bills paid up on ....

   b) report on work carried out on ....

   c) and a final report on ....

   A representative of the Embassy will have the right to visit the project at any time to make an evaluation of the project.

6. Other clauses

For the Embassy of	Person in charge of project	For the government of .... read and approved
Signature	Signature	Signature

Place and date: ....

FUNDS GOVERNED BY EMBASSIES (MISSIONS)
APPROVAL DOCUMENT OF A PROJECT

(complete all sections)

1. Country _____ Mission in charge _____

2. Title of project _____

3. Name of benificiary or organisation and person to be contacted
   _____

4. General information _____
   _____

5. Project objectives _____
   _____

6. Description _____
   _____

7. Cost a) Contribution - equipment   _____   _____ UC
                           - construction   _____   _____
                           - services   _____   _____
                           - administration   _____   _____

         b) local contribution   _____   _____

         c) other contributions (specify)   _____   _____

                               TOTAL   =====   =====

8. Payment profile:
   List of expenses -   January/March
                        April/June
                        July/September
                        October/December
                     TOTAL

9. Justification (basis for participation)

10. Comments/observations:

                       Recommended by   _____
                                         Person in charge

                       Approved by   _____
                                         Head of Mission

Copies to: General Manager or Controller
           Regional management
           Bilateral Programme Analysis Group

PROFILE OF A FINANCIAL REQUEST
FOR A RURAL DEVELOPMENT PROJECT

1. The Association - an introduction

In a maximum of 2 pages introduce the Association summarising the global activities already carried out and the results. Also demonstrate in a few lines that these results directly benefit the people who actively participated in carrying out the work.

2. Title of the project

Be sufficiently specific and indicate clearly the title of the project or programme for which the funds are required. The title should not be too long - 5 to 10 words, not more.

3. The Objectives

Differentiate between the global objectives and the specific objectives.

Avoid describing the global activities citing generalities such as "participate in the socio-economic development of the region". It is obvious and therefore unnecessary. On the other hand, "to introduce appropriate agricultural techniques in the region" can be a global objective, or even "to reduce the women's workload".

The specific objectives should be exact and detailed. For instance, to explain the global objective of "reducing the women's workload", select a specific objective such as "the digging of wells close to the living compounds".

The objective must be quantifiable and easily verified.

4. Justifying the Project

After stating the objectives, add a few paragraphs on the results of an environmental study carried out: problems raised, solutions to the identified requirements. Justify the choice of this project.

5. Organisation

There are three aspects which are vital for a donor to evaluate the feasibility of the project. These are the technical administrative and the participatory aspects of a project.

Therefore, submit concisely, but clearly, the technical organisation of the project, i.e. in the case of a mill:

- the make of the mill and motor
- place of purchase
- who will operate it
- who will maintain it
- working hours
- etc.

Also describe the proposed administrative organisation; i.e. who is in charge of the project administration (secretariat, finances, decisions, etc.)

Finally, in any project, there is a training and participation element that needs to be laid out in a proposal: what are the efforts to train the leader, persons in charge and the people.

6. The Resources

There are three kinds of relevant resources:

- The human resources (leaders, technicians, craftsmen).
- The material resources. Make out a list of materials available.
- Financial resources. At this stage, give an idea of the overall financial situation.

7. Implementation

Describe at length how the project is going to be launched and when it is expected to start (consider the agricultural time schedule for rural projects).

Enumerate the various operations, (the period of time needed to execute them), and submit everything in a global plan called "Plan of Implementation".

8. Costs (or budgets)

There are two kinds of costs. Investment costs and operating costs. Differentiate between these two categories of costs and do not mix them up.

Investment Costs deal only with purchase of material, equipment, machinery or other items which will be used over a period of one year.

Operating Costs relate to the current year and recur every year: salaries, oil, petrol, insecticide, maintenance costs, etc.

9. Profitability

In order to justify the project, indicate the economic and social benefits.

The economic benefits can be best judged by preparing an estimated Profit and Loss Account. Enter on one side, the total expenses for the year and on the other the annual income. If the income is higher than the expenses, then the project is profitable.

The operational costs as well as the annual replacement costs (depreciation) should be included in the Profit and Loss Account.

10. Financial Plan

Differentiate between four kinds of financing:

- firstly, calculate the <u>personal contribution</u> of the Association's members (cash or kind),
- secondly, if there is other local financing, add this too,
- thirdly, if the project manufactures saleable items, evaluate the <u>income from the sale of these products,</u>
- finally, the difference between the cost and the funds that can be mobilised locally (see above) gives the amount of the anticipated donor contribution which is being requested as foreign aid.

11. Request for External Finances

At the end of the proposal prepare on a single page the following cost calculation:

```
   Project Cost
   ─────────────────────────────────────────────
   -  less members' contribution
   -  less local financing
   -  less cash from sales
   ─────────────────────────────────────────────
   =  foreign aid required
   ─────────────────────────────────────────────
   -  less funding promised by another donor (if any)
   ─────────────────────────────────────────────
   =  funding requested from the donor to whom this
      proposal is being sent
   ─────────────────────────────────────────────
```

<u>Note</u>

Write down the amount in local currency and specify the official current rate of exchange between the local currency and the currency of the donor's country.

12. Annexes

If there are several annexes, summarise them on one page:

- Annual reports
- Technical reports
- Government recommendations
- Annual audited accounts
- etc.

LIST AND ADDRESSES OF DONORS

(not exhaustive)

1. Catholic NGOs

- MISEREOR, Mozartstrasse 9, Post Box 1450, 51 Aachen, Federal Republic of Germany
- Comité Catholique contre la Faim et pour le Développement (CCFD), 4 rue Jean Lantier, 75001 Paris, France
- CEBEMO, P.O. Box 90727, 2509 The Hague, The Netherlands
- Broederlijk Delen, Handelsstraat 73, 1040 Brussels, Belgium
- Development and Peace, 2110 rue Centre, Montreal H3K LJ5, Canada
- CARITAS, Lovenstrasse 3, 6002 Lucerne, Switzerland
- Action de Careme, Habsburgerstrasse 44, Postfach 754, 6002 Lucerne, Switzerland
- TROCAIRE (The Catholic Agency for World Development), 169 Bootertown Avenue, Dublin, Ireland
- Catholic Relief Service (CRS), 1011 First Avenue, New York, U.S.A.
- Catholic Fund for Overseas Development (CAFOD), 2 Garden Close, Stockwell Road, London SW9 9TY, UK

2. Protestant NGOs

- Bread for the World, Stafenbergestrasse 76, Stuttgart 1, Federal Republic of Germany
- HEKS/EPER, Post Box 168, Zurich, Switzerland
- CIMADE, Service Oecumenique d'Entraide, 176 rue de Grenelle, 75007, Paris, France
- Interchurch Co-ordination Committee for Development Projects (ICCO) Holland, P.O. Box 151, 3700 Ad Zeist, The Netherlands
- Evangelische Zentralstelle fur Entwicklungshilfe (EZE), Mittelstrasse 37, 5300 Bonn, Federal Republic of Germany
- Lutheran World Federation, 150 route de Ferney, 1211 Geneva 20, Switzerland
- Lutheran World Relief, 36 Park Avenue South, New York, N.Y. 10010, USA
- The United Methodist Church, 475 Riverside Drive, New York, N.Y. 10115 U.S.A.
- Christian Aid, 240/250 Ferndale Road, Brixton, P.O. Box 1, London SW9 8BH, UK
- Interchurch Fund for International Development (ICFID), Suite 314, Toronto, Ontario M4T 1M8, Canada

3. Non-religious NGOs

- OXFAM-UK, 274 Branbury Road, Oxford OX2 7DZ, UK
- OXFAM BELGIUM, 89 rue de Consert, 1050 Brussels, Belgium
- OXFAM-US, 111 Broadway, Boston, Massachusetts, U.S.A.
- Euro-Action-Accord, Francis House, Francis Street, London SW1 1DQ, UK
- Deutscher Volkshochschul-Verband E.v. (DVV), Rheinallee 1, 5300 Bonn 2, Federal Republic of Germany
- Terre des Hommes, 26 rue des Bateliers, 93400 St. Ouen, France
- NOVIB, Alrastraat 5-7, The Hague, The Netherlands
- HIVOS, Beeklaan 387, The Hague, The Netherlands
- National Committee for Co-operation for Development (CNCD), 76 rue de Laeken, 1000 Brussels, Belgium
- Helvetas, St. Moritzstrasse 15, 8000 Zurich, Switzerland
- Private Agencies Collaboration Together (PACT), 777, United Nations Plaza, New York N.Y., 10017, U.S.A.
- A.T. International, 1311 H Street, N.W. 1200, Washington D.C. 20005, U.S.A.
- Campagne Française contre la Faim, 42 rue de Cambronne, 75015, Paris, France
- S.O.S. Faim, 4 rue Laine, 1000 Brussels, Belgium
- Frères des Hommes, 20 rue de Refuge, 78000 Versailles, France
- Swissaid, Jubilaümstrasse 60, 2000 Berne, Switzerland
- Interaction, 2101 L St., N.W. 916, Washington D.C. 20037, U.S.A.
- German Agro-Action, Adenauerallee 134, 5300 Bonn 1, Federal Republic of Germany
- Liaison Committee of NGOs for the Environment (ELC) P.O. Box 72461, Nairobi, Kenya
- AFRICARE, 660 First Avenue, New York, N.Y. 10016, USA
- War on Want, 467 Caledonian Road, London N7 9BE, UK
- USA for Africa, 6151 WW. Century Blvd, Los Angeles, CA 90045, USA
- Technoserve, 11 Belden Avenue, Norwalk, Conn 06852, USA

4. Foundations

- The Ford Foundation:
  Headquarters, 340 East, 43rd Street, New York, N.Y. 1007, U.S.A.

  East and Southern Africa, P.O. Box 41081, Nairobi, Kenya

  Asia :
  . Ford Foundation, 55 Lodi Estate, New Delhi, India
  . Ford Foundation, Jalan Tamankebon, Sirish I/4, Jakarta Pusat, Indonesia

- The Commonwealth Foundation, Marlborough House, Pall Mall, London SW1Y 5HU, UK

- The Rockerfeller Foundation, 1133 Avenue of the Americas, New York, N.Y. 10036 U.S.A.

- Aga Khan Foundation, 7 rue Versonnex, Case Postale 435, 1221 Geneva 20, Switzerland

- African Development Foundation, 1724 Massachusetts Avenue N.W., Suite 200, Washington D.C. 20036, U.S.A.

German Political Party Foundations

- Friedrich Ebert Foundation, Kölnerstrasse 149, 5300 Bonn Bad Godersberg

- Korad Adenauer Foundation, Rathausallee 12, 5205 St Augustin 1b/Bonn

- Institut für Internationale Solidaritat (ISI), Rathausallee 12, Postfach 1260, 5205 St. Augustin 1, bei Bonn

- Freidrich Nauman Foundation, Margarethenhof, Konigswinterer Str. 2-4, 5330 Konigswinter 51

5. Bi-lateral Aid

- Technical Co-operation Service of the Ministry of Foreign Affairs, Muzenstraat 30, The Hague, The Netherlands

- Co-operation for Development and Humanitarian Aid (DDA), 73 Eigerstrasse, 3003 Berne, Switzerland

- Ministry for Co-operation, 20 rue Monsieur, 75007 Paris, France

- Overseas Development Ministry (ODM), London SW1E 5DH, UK

- Deutsche Gesellschaft für Technische Zusammenarbeit (G.T.Z.), Dag-Hammarskjöld-Weg 1, P.O. Box 6236 Eschborn, Federal Republic of Germany

- Canadian International Development Agency (CIDA), 200 Promenade du Portage, Hull, Quebec K1A O64, Canada

- United States International Development Co-operation Agency (US-AID), Washington D.C. 20523, U.S.A.

- Administration Générale de la Coopération au Developpement (AGCD), 5 place du Champ de Mars, B.P. 57, 1050 Brussels, Belgium

- Dipartimento per la Co-operazione allo Sviluppo, Ministry of Foreign Affairs, via Salvadore Contavini 25, 00194, Romè, Italy

- Swedish International Development Authority (SIDA), Fregattragen, 18500 Stockholm, Sweden

- DANIDA, Ministry of Foreign Affairs, Amallegade 7, Copenhagen, Denmark

- Norwegian Agency for International Development (NORAD), P.O. Box 8142, Oslo 1, Norway

- Finnish International Development Agency (FINNIDA), Ministry for Foreign Affairs, Mannerheimintie 15C, 00260 Helsinki 26, Finland

- Australian Development Assistance Bureau, Ministry of Foreign Affairs, P.O. Box 887, Canberra City 2601, Australia

- Ministry of Foreign Affairs, Technical Cooperation Service, Wellington, New Zealand

- Economic Cooperation Bureau, Ministry of Foreign Affairs, N°2-2 Chome Kasumi Gaseki, Chiyoda-ku, Japan

Regional Organisations

- Commonwealth Fund for Technical Co-operation (CFTC) Marlborough House, Pall Mall, London SW1Y 5HX, Great Britain

- Agence de Coopération Culturelle et Technique (ACCT), 19 Avenue Messire, 75008 Paris, France

- Commission des Communautés Européennes (C.E.E.), 200 rue de la Loi, 1049 Brussels, Belgium

- Centre de Recherche pour le Developpement International (CRDI), P.O. Box 8500, 60 Queen St., Ottawa K1G 3H9, Canada

- SAREC, Birgir Jarlsgaten 61, S-105 25, Stockholm, Sweden

Some of the above agencies have representatives in different developing countries. Their local addresses - if any - could be obtained from the relevant Embassies.

6. **U.N.O./Multilateral Aid**

- United Nations Development Program (UNDP), 866 United Nations Plaza, New York, N.Y. 10017, U.S.A.

- International Labour Organisation (ILO), 1211 Geneva 22, Switzerland

- World Health Organisation (WHO), 1211 Geneva 27, Switzerland

- UNESCO, 7 place de Fontenoy, 75700 Paris, France

- UNICEF, 866 United Nations Plaza, New York, N.Y. 10017, U.S.A.

- Food and Agricultural Organisation of the United Nations (FAO), via delle Terme di Caracella, 00100 Rome, Italy

- United Nations Environmental Program (UNEP), P.O. Box 30552, Nairobi, Kenya

plus their representatives in your country.

7. **Development Banks**

- World Bank, 1818 H Street, N.W. Washington D.C. 20433, U.S.A.

- African Development Bank, P.O. Box 1387, Abijan, Ivory Coast

- Asian Development Bank, 2330 Roxas Boulevard, Metro Manilla, The Philippines

- Inter-American Development Bank (IDB), 808, 17th Street, N.W., Washington D.C. 20577, U.S.A.

plus the national development banks of individual countries.

TECHNICAL CHARACTERISTICS
OF FLEXIBLE FUNDS

An example of the SIX'S Association -
"Se Servir de la Saison Sèche en Savanne et dans le Sahel"
(Making use of the dry season in the Savanne and the Sahel)

1. OBJECTIVES

Making financial resources available to formal or non-formal groups working at village level, to be used as grants or loans.

The use of these resources for co-financing activities, together with the resources of these local groups. Activities: which were identified at the time the funds were given, selected and identified by the requesting group itself receiving priority for using the beneficiary's own resources.

2. METHOD OF MANAGEMENT OF FUNDS

a) Structured at three levels

SIX'S - International: At the Annual General Assembly of the Association, decisions are taken regarding the distribution of the funds. The Executive Secretary takes on the functions of motivation and control.

SIX'S - District: A district is a geographical area where a certain number of peasant groups work. A District Committee, comprising only managers of group unions, manages the funds allocated to it annually by the General Body.

Grassroots level: Each group working in the territory of a district can obtain a subsidy or a loan on negotiation with the Committee of the District.

b) The annual funds are obtained at the SIX'S International level through funding agreements with various private and public donor organisations. Generally the agreements cover a period of 2 or 3 dry seasons. Contributions from various agencies are managed in common; no single contribution being allocated for a specific region or for a specific use. A common Annual Report justifies in common the use of all the funds given by the different donor agencies. Each donor contribution is not justified separately.

c) The method used for distributing annual funds managed at the international level is the following:

At every SIX'S General Assembly (in October or just before the dry season), the members divide the total foreign aid (for the dry season) according to the following basis of distribution in two stages:

e.g. 10 million UC

1st Stage: 13 % to general expenses ...... 1.3 million UC

87 % to districts .... 8.7 million UC

= 10.0 million UC

2nd Stage: This 8.7 million is divided among:

a) the existing districts depending on the age of each district organiation, e.g.

- over 5 years old = 0.6 million
  4 years old = 1.0 million
  3 years old = 1.5 million
  2 years old = 0.8 million
  1 year old = 0.5 million

b) The new districts to be created during the next dry season: e.g. 0.1 to 0.2 million per new zone.

d) Once the money has been deposited in the districts accounts, it is the District Committee which is solely responsible for its management without interference from the General Secretariat (except when it comes to helping out with accounting training and follow-up).

During the dry season money is disbursed to the districts as funds come in from donor agencies. In some years the amount is less then expected (delay in payment from an agency, for instance). The money deposited after the 31st May is cancelled and is not carried forward in the distribution for the next dry season.

3. ANNUAL SCHEDULE OF MANAGEMENT OPERATIONS

The distribution of funds is done annually and can be obtained only on presenting the previous dry season's accounts, certified by an independent Chartered Accountant. Everything should, therefore, be "tied-up" as fast as possible. The schedule is as follows:

31st May   -   entering of accounts for the previous dry
                    season is stopped.

June   -   audited statements of accounts are fed to the
                    computer, district by district and for the entire
                    SIX'S.

Jul/Aug/Sep   -   reports are completed on the utilisation of funds
                    (per district, per country and for SIX'S Inter-
                    national) by the district heads, national coordi-
                    nators and the Executive Secretary General of
                    SIX'S.

September   -   the financial report of the accountant is prepa-
                    red.

October   -   inspection of the working report of the prece-
                    eding dry season and the financial report of the
                    accountant by the SIX'S General Committee,
                    which decides on the distribution per district to
                    be made in the next dry season.

December   -   remittance of first instalment of annual funds to
                    each of the bank accounts of the districts (one
                    bank account per district).

Jan to May   -   allocation of co-finance to the groups by each
                    district committee SIX'S.

Jan to April   -   remittance of second and third instalments of
                    the annual funds.

31st May   -   end of expenses for the current dry season.
                    The accounts are from 1st June to 31st May.

4. MANAGEMENT METHODS AT A DISTRICT LEVEL

Are the district programmes prepared for the entire dry season?

Every year at the beginning of the dry season, those in charge of
a SIX'S district know the volume of foreign aid that SIX'S intends
to.distribute in the next 6 months. Very few of them will establish
programmes at this stage; most of them will decide on the use of
funds when the money arrives. They have been disappointed
earlier when the expected contributions were not sent for one
reason or another, or delayed. As these forecasts were worked out
with the co-operation of the groups, everyone was disappointed.
Therefore, district committees wait for an instalment of the funds
to be deposited in the bank account of the district before proceed-
ing to programme its utilisation.

Who decides on the utilisation of the funds?

Each group submits its request to the district management committee or sometimes to the General Committee if it has been conveyed that everything has to be decided upon in the presence of all concerned. A request can comprise varied activities provided that the person requesting gives the assurance that he will bring in his own resources to make this operation a success. The total finances requested is often higher than the amount available.

If there is a choice, the decision is made by the persons selected by the groups (i.e. those who form the committee). They consider the following criteria: volume of own resources; number of persons interested; possibility of obtaining and repaying an internal loan; risks involved; the possibility of making this work out without delay.

The primary groups have a tendency to invest in operations with an economic focus , of which the expected results could satisfy several kinds of urgent requirements and gradually cover their own social costs. In most cases, the unions of groups reserve a part of the amount received to cover management costs.

To whom do they lend?

Except in the case of individual loans for improving living conditions, the committees lend only to primary groups, to facilitate management, to reinforce inter-dependency and to ensure a collective guarantee at the same time. The group is free to use the credit to lend to its members or for a collective investment. Often the two options are combined; for instance, the cost of fencing a collective field is met by the group which values it, whereas the watering cans will be repaid by each member who would have received one to cultivate some vegetable beds.

Whom do they repay?

The individual repays his group which can put pressure on him if and when needed so that the group itself will be able to repay the District Committee. It is often advantageous to purchase the requirements (wire, seeds, etc.) wholesale. That is why a group receives "in kind" and repays "in cash".

Should the interest rates be high or low and the duration of payment fixed or flexible?

For the District Committee which receives a grant which it can lend to its groups, it is tempting to adopt soft repayment conditions. However, an interest rate lower than currently operative in official credit institutions will be a source of contradiction with the credit norms of the public sector. A repayment by way of kind (a bull, two years after obtaining a loan for a calf) is sometimes used, but it can cause auditing problems. SIX'S International allows each district the freedom to decide by itself the interest rates and methods of repayment.

As for delays in repayment, if the District Committee takes out a guarantee against risks (e.g. livestock mortality), it can specify a definite period of repayment; if it does not wish to or cannot obtain this guarantee (e.g. risk of drought in the case of a loan intended for marketable production), the borrower should repay even if he has not harvested anything, a practice which is contrary to principles of aid. That is why in cases of disaster, District Committees prefer to carry forward to a better year all or part of the repayment.

Who keeps the accounts?

The documentary proof of resources spent is kept by the beneficiaries themselves; e.g. co-financing of multiple groups could be aggregated and audited at the district level.

A simple system of entering accounts has been adopted for all SIX'S accounts' establishments (being 60 in 1986); it enables the feeding of the results into the computer and their distribution according to the nature of expenses and according to the type of activity.

What does district accounting mean?

Where local groups are concerned, the local contributions of beneficiary groups (membership fees, working days, repayment of loans) are more important at times than are foreign aid contributions. The accounting system will show if their own efforts require foreign contributions. Those who give aid also will want to see that there is a synchronisation of local effort with foreign aid.

Finally, the expense statement alone is inadequate: certainly it allows estimating investment proportions, and helps to analyse the importance of expense categories (transport, labour, administration) but gives little importance to the real beneficiary. That is why a statement of effort contributed is sought for each village, including the resources utilized, their source (local contribution or aid) and the end result for each village, each year.

PEASANTS OF SAHEL INNOVATE A NEW INTERNATIONAL AID SYSTEM

This is an innovation of the NAAM groups of Yatenga in Burkina Faso.

The Mossis woman spends most of her time attending to three tasks: fetching water, working in the fields and cooking meals/pounding cereal. During intensive working periods the time needed for pounding cereal reduces the time she could spend in the field. Her days are long and she is very tired.

Taking into consideration these facts, some NAAM group leaders and SIX'S leaders, with the financial aid of the Ecumenical Council of Churches or UNICEF, bought mills to pound cereal at a cheap price. These modern mills run on a petrol motor, are regularly maintained by the local mechanics while major repairs are carried out in Ouhigouya, capital of Yatenga.

Each mill is run by a Management Committee of 12 members: six men and six women. The responsibilities are distributed among as many persons as possible so as to make them participate. For instance there are two treasurers: an elder to keep the cash and a young person to keep the books of accounts. The miller is an adolescent who works with a young friend.

These mills which are greatly appreciated by the women have been given by the Ecumenical Council of Churches or some other donor agency.

One of the important lessons experienced by the Mossis is that of solidarity. "You should not always only help yourself but you should also help others..."

This fundamental truth is used by NAAM group leaders. Solidarity and inter-dependence requires those who receive a mill to repay the purchase price in a way that another group could in their turn get a mill for themselves.

A member/woman of the Management Committee of a mill innovated the concepts of "Father Mill", "Daughter Mill", and "Son Mill" to describe an economic system through which the mill was going to "make small ones" and reproduce itself.

The "Father Mill" is the ancestor, the founder of the "family" which he generates; it is he whom you respect and consult. The first child of the "Father Mill" (and Mother) will be a "daughter". It will be called "Daughter Mill", just as a daughter leaves the village at marriage to join another family and give herself to her husband to have children. The "Daughter Mill" will be the first expression of inter-dependence and of the good management of the "Father Mill" group. After 2 years the mill would have made enough profit to give birth to another "daughter".

The "Father Mill" will age; without a son it will not have descendants and the family will come to an end; it will then be likened to a "cur-se", as is the case of a father who has no son. The "Father Mill" (and Mother) should, therefore, give birth to a son who will be capable of replacing the father when he will be old, tired and worn out. The "Son Mill" will assure the replacement of the old mill and the continuity of inter-dependence. In his turn, he will learn to "help the others": and the system of replacement and replication goes on.

This system is carried out in several villages. The grants received through international aid are in fact repayed 100% by the groups. Inter-dependence takes place. The social pressure of those who await a new mill (like the first ones) induce the others to pay up sooner than expected.

I have noticed these facts. I have carefully read the accounts kept by the Management Committees of Mills. I have admired the way peasants have innovated an economic system based on their own traditions and helped more people gain by the efforts of others.

I wonder if these Sahelian peasants and their group leaders have not in fact innovated the principles for a new aid system. Can we not ask the privileged rich countries of the North, not to lend money to Third World countries, but to offer grants which, when re-distributed and loaned to organised groups will, in their turn "generate resources for investments by other groups".

This obviously means that the organisations of the North and of the South would have to meet and discuss the mechanisms which could help to set up a new aid system based on the innovative developmental strategies of the Sahelian peasants.

The North could grant un-allocated flexible funds to a common fund managed by peasants' organisations who would not have to repay any of it. The beneficiary organisations could then lend this money to local groups on condition that they agree to repay the amount to a neigh-bouring group. Such an aid system would obviously mean that the local partners agree to repay. Of course, the risk is great but it would be preferable to the wastage of grants and loans seen in the present aid system which, in most cases, does not entail the active participation of the local people in their own development.

The innovative Sahelian approach shows how development could be based on group responsibility towards other groups which in turn could generate a multiplication of efforts.

Traditional economists certainly have a lesson to learn from these Sahelian peasants who knew how to integrate the theories of investment and amortization with the values of INTER-DEPENDENCE and SOLI-DARITY.

Fernand Vincent, January 1982.

ANNEXES PART II

SAVINGS AND CREDIT SYSTEMS

SPECIMEN PROMISSORY NOTE

1. Capital sum borrowed:

   13,000 Units of Currency

2. Rate of Interest per centum:

   10 %   Penal rate: 14 %

3. Grace period:

   2 months

4. Date borrowed: 21 December 1987

I, the undersigned _____Yusuf Ismail_____
                              (name)

of ___Penang Development Association_____   promise to
                (address)

pay to __Malaysian Credit and Savings Bank, Penang,___ the sum of
           (name of lending institution)

Thirteen thousand (Units of Currency)   ___(13,000 UC)__
(amount in words)   (amount in figures)

in thirteen (13) instalments of 1,000 Units of Currency per month

with effect from 1 February 1988.

STAMP

Signature of borrower

SPECIMEN PERSONAL GUARANTEE

To:

Malaysian Credit & Savings Bank,
Penang

In consideration of your lending the sum of _____ 10,000 _____ UC to Joseph Chin of the Penang Development Association, we, the undersigned, Liu Chiang and G.M. Singh of the Penang Development Association undertake to repay on demand the said loan with interest thereon at the rate of ten per centum (10 %) per annum unconditionally, in case of default on thé part of the borrower.

Signature/Thumb print   Signature/Thumb print

(Liu Chiang)   (G.M. Singh)

1. Witness

         ....
         Signature

         ....
         Name and address

2. Witness

         ....
         Signature

         ....
         Name and address

DATE: 10.8.88

RAFAD

RECHERCHES ET APPLICATIONS DE FINANCEMENTS ALTERNATIFS AU DÉVELOPPEMENT
RESEARCH AND APPLICATIONS FOR ALTERNATIVE FINANCING FOR DEVELOPMENT
INVESTIGACIÓN Y APLICACIÓN DE FINANCIAMIENTOS ALTERNATIVOS AL DESARROLLO

List of information to be supplied
when applying for RAFAD Guarantee funds

This document is intended as a guide to the questions to be asked and the replies to be obtained in order to identify a RAFAD Guarantee. This work can be carried out by a support organisation, a local consultant or any other person nominated to do so.

This document could possibly be sent to the local group, if it is capable of answering the questions on its own.

We therefore ask that a report be prepared which allows for a maximum of information to be obtained. Do not hesitate to give any additional information which could prove useful or to adapt this document to the real context of the future RAFAD partner. Please also enclose a copy of any documents which complete the dossier: by-laws, balance sheet and accounts, reports, etc.

I. IDENTITY

1. Name of the group

2. Full address

3. Name and title of the group leader

4. Legal status

II. PROFILE OF THE GROUP/PARTNER

1. Activities

   - Describe briefly the activities of the group concerned.

   - Try to evaluate the quality and impact of the activities.

   - Try to find out the activities which the group wants to develop in the future.

   It would be helpful to have some progress reports, programmes, projects or any other documents concerning these activities.

2. Financial situation

- Determine the resources available at the bank or in cash.
- Draw up an inventory of the group's assets: land, buildings, machines, furniture, vehicles, etc.
- Can you estimate the total sum of annual income from the sale of the group's production?
- If the group has debts, is it possible to know the reason for borrowing and how the loan was utilised?
- Try to evaluate the group's present financial situation. Enclose, if possible, the income and expenditure accounts and the balance sheet.

3. Organisation and participation

- Introduce the group: origin, initial idea, members, responsibilities, qualifications, competence, etc.
- Describe how the group organised itself to carry out its projects.
- Describe the different ways in which the members participate in the management and working of the group's projects.
- If training activities for members exist, decribe them.

4. Savings/credit

- If the members of the group have begun to save, individually or collectively, describe how they have done so and the results of this saving.
- If a group has already obtained a credit or a loan, explain what, how and from whom it was obtained; what the interest was and how the loan was repaid.
- If relevant, it would be useful to know the name of the bank which granted the credit so that we know which banks lend to those who have limited resources.

5. Relations

- The group most probably has relations with the local government and its technical services. If relevant, describe the content of such relations and comment on their usefulness.
- If the group receives or has received external or local financial help, please indicate how much, from whom and how.
- It is important to know if the group is in contact with other NGOs, associations or partners and networks. Describe these relations if they exist.

III. CREATION OF A RAFAD GUARANTEE

1. Describe as fully as possible the reasons that have led the group to request a RAFAD Guarantee.

2. Try to describe concretely how the RAFAD Guarantee will work. Give details; for example, who will be the beneficiaries, how they will be chosen, what kind of loan they would receive and why, as well as the proposed repayment period and interest.

3. If the group has already thought about organising the management of its funds, please indicate:

   - who will be responsible for the management of the funds and who will take the decisions.

   - how you will organise the system of loans in such a way as to have the repayments made entirely within the time limits (use of solidarity and the use of social pressure by one group of individuals on another group etc.).

4. RAFAD Guarantees do not insist on a project proposal. It is, however, useful to know the activities to which the group intends to attribute the funds lent by the bank.

5. Try to supply the following technical information and data for the creation of a RAFAD Guarantee:

   - total amount of the guarantee requested (justify)

   - for how long?

   - name and address of the bank with which the group expects to negotiate its guarantee to obtain credit and to which the RAFAD Foundation will send the bank's letter of credit/-guarantee

   - what is the rate of interest which you expect to negotiate or which has been obtained?

   - will the bank loan you, in local currency, an amount lower, equal or higher than the value of RAFAD letter of credit?

   - what are the expected repayment terms of the guarantee?

6. Does the group need support in management and organisation? If so, what type of support would you like, from whom, and why? Does the group already receive technical assistance? What type and from whom?

7. Has the group fixed objectives to be attained by asking for a RAFAD Guarantee? If so, try to describe them.

8. The RAFAD Guarantee will <u>facilitate</u> the obtaining of a credit or an overdraft, which must be reimbursed with interest. Will the activities developed with this credit allow for reimbursement leaving a surplus? Give details of how this has been envisaged.

9. The group may well have some difficulties in replying to these questions and in managing its guarantee funds. Could you indicate what difficulties you anticipate and how you would overcome them.

10. This document/questionnaire cannot fit all possible situations. You should, therefore, make any additional remarks which could help us to understand your particular situation and make an efficient decision.

    Collaboration between RAFAD and the groups which use its guarantees is based on <u>mutual confidence</u>.

ANNEXES PART III

SIMPLE ACCOUNTS, BUDGETS AND CASH-FLOWS

SPECIMEN RECEIPT

Payment made by the Association

RECEIPT NO: *127*

Received from:

Rural Development Association of

Bambug District

the sum of _____*one hundred*_____ UC _____*100,00*_____
(amount in words)   (amount in figures)

being payment for: _____

_____*advance for mill*_____

_____

_____

Date: *15/6/88*   STAMP

                                   Signature
                                   Juderson

- Original for the accounts of the Association

- Copy for the person who receives the cash

SPECIMEN RECEIPT
Acknowledging payment made to the Association

RECEIPT No. 324

The Rural Development Association of Bandang District acknow-
ledges with thanks the sum of _hundred_ (_100,00_) UC
                       (amount in words)   in figures

from :

Name: _S. Tudor_

Name of Association: _ACEPA_

being for: _Saving deposit_
              (purpose of payment)

Date: _12/7/87_   _Judersons_
                                     (signature)
                            Accountant/Treasurer
                            R.D. Association of
                            Bandang District

- Original to the person who pays cash

- A copy for the Cash accounting of the Association

When it is not possible to obtain a receipt for the purchase of certain items or from certain shops, make out the following document which should be filled in by hand by the person who bought the goods on behalf of the Association.

SPECIMEN CASH VOUCHER

Foundation for Integrated
Rural Development (FIRD)
Bangra
Bangkok

Quantity	Particulars	Unit cost	Total cost
20 kg	Sugar	12 uc per kg	240 UC
		TOTAL	240 UC

Signature of a witness
if the amount exceeds
1,000 UC

Signature

Date: 12/8/88

OPENING A BANK ACCOUNT

To open a Bank Account, write the following letter to the Bank:

Vinivida NGO Coalition
Angunawila
Mundel

15th June 1988

The Manager,
Sampath Bank,
Chilaw Branch,
P.O. Box 1234
CHILAW

Dear Sir,

OPENING A BANK ACCOUNT

Please be good enough to open a Current Account in your Bank in the name of VINIVIDA NGO COALITION.

Please find enclosed the following documents:

- The Constitution of the Association and its Registration Certificate from the Ministry of Rural Development.
- Minutes of the General Meeting nominating the persons authorised to sign on behalf of the Association.
- A specimen signature card duly filled in.

We are at your disposal for any further information that you may need.

Yours faithfully,

Vinivida NGO Coalition

....
K.D. Pieris   C.B. Fernando
Accountant   Chairman

Encl:
- Constitution of the Association and its Registration Certificate from the Ministry of Rural Development.
- Minutes of the General Meeting nominating persons to sign on behalf of the Association.
- Specimen signature card (obtain this card from the bank earlier)

CHEQUES

1. CROSSED CHEQUES

Used most often. The safest.

CAUTION: It can never be cashed by the beneficiary and can only be deposited by the beneficiary in his bank and credited to his account. It is only after this that he can withdraw the money.

```
| COUNTERFOIL            | BANK OF KARNATAKA, P.O. Box 112, Bangalore        |
|                        |                                                   |
| Date: 25.03.88         |                             Date: 25.03.88        |
|                        |                                                   |
| Payee: P. Sharma       | Pay to _____ Mr P. Sharma _____     |
| For: Supply for        | Rupees _____ One thousand only _____    |
|      Stationery        | _____ Rs. 1,000/-|
|                        |                                                   |
| Bal.b/f:    60,000/-   | Kanakapura Young Farmers Club                     |
| This cheque: 1,000/-   | Account No. 1272987-001                           |
| Total balance: 59,000/-|                   S. Kanakaraj        S.P. Joshi    |
|                        |                   ...........         ...........  |
|                        |                   Chairman            Treasurer    |
| "094171"               | "094171"                                          |
```

2. NOMINATIVE CHEQUES

- Bears the name of the person to whom you wish to pay.
- Should not be crossed. It can be cashed only by the beneficiary (on producing his identity card).

```
| COUNTERFOIL            | BANK OF KARNATAKA, P.O. Box 112, Bangalore        |
|                        |                                                   |
| Date: 27.03.88         |                             Date: 27.03.88        |
|                        |                                                   |
| Payee: P.K.Roy         | Pay to _____ Mr P.K. Roy _____     |
| For: Loan advance      | Rupees _____ Two hundred only _____    |
|                        | _____ Rs.  200/-|
|                        |                                                   |
| Bal.b/f:    59,000/-   | Kanakapura Young Farmers Club                     |
| This cheque:  200/-    | Account No. 1272987-001                           |
| Total balance: 58,800/-|                   S. Kanakaraj        S.P. Joshi    |
|                        |                   ...........         ...........  |
|                        |                   Chairman            Treasurer    |
| "094172"               | "094172"                                          |
```

3.   BEARER CHEQUES

CAUTION: A bearer cheque can be cashed by anyone if it is lost. It has the same value as a currency note.

COUNTERFOIL	BANK OF KARNATAKA, P.O. Box 112, Bangalore
Date: 28.03.88	Date: 28.03.88

Payee: Cash (K. Menike)
For: Sale of 100 kg
    peanuts to Co-op
    Shops
Bal.b/f: 58,800/-
This cheque: 500/-
Total balance: 58,300/-
	Pay _____ CASH _____ to the bearer
	Rupees _____ Five hundred only _____
	_____ Rs. 500/-
	Kanakapura Young Farmers Club
	Account No. 1272987-001
	              S. Kanakaraj S.P. Joshi
	              ••••••••••• •••••••••••
	              Chairman Treasurer

"094173"	"094173"

SPECIMEN OF A DEBIT NOTE

THAI PEOPLE'S BANK

POST BOX 112 - BANGKOK

FOUNDATION FOR INTEGRATED RURAL DEVELOPMENT
ACCOUNT NO..........   Date..........
                                          Ref..........

CURRENT ACCOUNT

Kindly note that your account has been **DEBITED** with

_____ on _____ details of
       (UC)   (date)

which are as follows:

Payment of your Invoice No. 1334 dated 31.3.88

Gross	Commission	Handling charges	
3000	1 % (30)	-	3,030

Yours faithfully,

THAI PEOPLES BANK

p.p. Manager

If you have any queries regarding the above transaction
please address your inquiries to the Operations Manager,
quoting your Account Number and this transaction reference.

SPECIMEN OF A CREDIT NOTE

THAI PEOPLE'S BANK

POST BOX 112 - BANGKOK

FOUNDATION FOR INTEGRATED RURAL DEVELOPMENT
ACCOUNT NO.........   Date..........
                                        Ref.........

CURRENT ACCOUNT

Kindly note that your account has been **CREDITED** with

_____ on _____ details of
(UC)   (date)

which are as follows:

From EcoSolidar (Switzerland) for 3,000 Swiss francs

Gross	Commission	Rate of exchange	
61,500	1 %	at 20.50	60,885 UC

Yours faithfully,

THAI PEOPLES BANK

p.p. Manager

If you have any queries regarding the above transaction
please address your inquiries to the Operations Manager,
quoting your Account Number and this transaction reference.

SPECIMEN OF A TRANSFER ORDER

KANAKAPURA RURAL DEVELOPMENT ASSOCIATION

89th Milepost,
Kanakapura

15th January 1988

The Manager of the Bank of Karnataka
P.O. Box 112, Bangalore

Dear Sir,

TRANSFER ORDER

Please be good enough to transfer by debit of our account No. 127/29/786-001 the sum of 3,500 UC (Three Thousand Five Hundred Units of Currency)

to the order of: The Cooperative Society
M.G. Road
Bangalore

stating: in settlement of your Invoice No. 1275
of 15th June 1988.

Yours faithfully,
KANAKAPURA RURAL
DEVELOPMENT ASSOCIATION

Ravi Shekhar,
Chairman

SPECIMEN ORDER FORM

Use order forms for confirmation of
order and guarantee of payment.

YOUNG FARMERS LEAGUE
Jalan Samat
Jogjakarta

ORDER FORM No. 2539

CBA (Pvt.) Ltd.
Mari Street
Jogjakarta

Please deliver the following items:

UC

Quantity	Description	Unit Price	Total Net price	Remarks
24	Ball point pens	3.50	84.00	Colour: black or blue
2	Packets of A4 size paper	75.00	150.00	
1	Box of Stencils	50.00	50.00	

Remittance: payment in 30 days

Delivery: immediately if available

By Sudarsons
Accountant

Date: 23 May 1988

When goods are delivered without being paid for, a "goods delivered" voucher is sent along.

Do the same for your customers.

<u>Note</u>: Do not forget to invoice them for the goods in the next few days.

SPECIMEN GOODS DELIVERED VOUCHER

CBA (Pvt.) Ltd.
Mari Street
Jogjakarta

GOODS DELIVERY VOUCHER NO. 00454

To: Young Farmers League, Jalan Samat, Jogjakarta

UC

Quantity	Description	Unit Price	Total Price
24	Ballpoint pens	3.50	84.00
2	Packets of A4 paper	75.00	150.00
1	Box of Stencils	55.00	55.00

Date: 23rd May 1988

. . . .
Signature

SPECIMEN OF INVOICE

Ceylon Xerox Ltd.,   Date : 25.05.87
P.O. Box 327   Ref. : 675
Colombo 2

┌─────────────────────┐
│ INVOICE N° 1234   │
└─────────────────────┘

To: Wilpotha Development Association   Customer's Code : 34001
    89th Milepost
    Wilpotha

Order ref.	Description	Quantity	Unit Rate	Amount
1001	Toner	1	750/-	750.00
0002	A4 W. paper (70 G.S.M.)	3	155/-	465.00
1046	A4 W. paper (80 G.S.M)	3	195/-	585.00

Stamp ....

Signature ....

EXAMPLE OF KM COST CALCULATION
Toyota Hiace Van

- Vehicle purchased on 12.06.87 in Jakarta for 175,000 UC
- Depreciation over 5 years
- Vehicle regularly maintained by Royal Garage, Jakarta Pusat
- Km run from July 1987 to 30th June 1988:

Km on 30.6.88	16,912
Km on 1.7.87	2,512
Km run during year =	14,400

COST CALCULATION PER KM

in UC

1. Depreciation:

$$\frac{\text{Purchase Price}}{\text{Depreciation Period}} = \frac{175,000}{5 \text{ years}} = 35,000$$

2. Fuel Consumption

   As per Monthly Statements for the year = 11,500

3. Maintenance Costs

   As per invoices for the year: repairs, service costs, spare parts, tyres, etc. = 3,800

4. Miscellaneous Expenditure

   Insurance, taxes, a year = 1,000

   TOTAL COST = 51,300
   ======

RESULT

$$\text{Price per Km} = \frac{\text{Annual cost}}{\text{No. of Km run}} = \frac{51,300}{14,400} = 3.56 \text{ UC}$$

In conclusion: each km run with the TOYOTA HIACE VAN costs 3.56 UC (approx)

ANNEXES PART IV

MANAGEMENT
OF
SMALL PROJECTS

DETERMINING THE ESTIMATED STOCK OF A CEREAL BANK

(during the dry season)

Consider various indicative factors:

- the number of families in a village

- the daily cereal consumption per family

- the probable duration of the dry season

- the number of families who have requested the assistance of a cereal bank in previous years

Exercise:

In a Javanese Village, there are about 150 families:

- each family consumes an average of 2 kg cereal per day;

- the dry season is from March to August (about 180 days);

- in 1987, 60 families requested the assistance of the cereal bank.

With this information:

- determine the cereal comsumption of the village during the dry season

- determine the estimated stock which should be kept in the cereal bank

Solution:

1. The daily cereal comsumption for the village population during the
    dry season is:

    2 kg X 150 familes = 300 kg cereal per day

2. The probable cereal comsumption for the village population during
    the dry season is:

    300 kg cereal X 180 days = 54,000 kg

3. Only 60 families regularly requested assistance from the cereal
    bank last year. Therefore, it is assumed that the other families
    are self-sufficient during the whole year.

    Consequently the estimated stocks should be:

    2kg/day/family X 60 families X 180 days = 21,600 kg.

    Add a safety margin of 10 % to cover various festivals and contin-
    gencies:
    21,600 kg + 2,160 kg = 23,760 kg

Conclusion:

The group will need 23,760 kg of cereal to cover the village needs
during the dry season.

CALCULATION OF PROFIT FOR A RICE MILL

How many kilos of rice should be milled per month or per day for a mill to be profitable?

1. Some facts:

   - Cost of Mill = 150,00 UC

   - 1 litre of diesel = 100 kg of rice milled

   - The service return of a kg of milled rice is 1 UC

   - Machine needs oil change after every 10,000 kg milled = 5 litre of engine oil.

   - Depreciation over 5 years

   - Consumption of diesel per year: 1,000 litres at 7 UC

2. Calculating the profit for one year

   Business turnover

   100,000 kg X 1 UC   =   100,000

   Variable Costs

Diesel 1,000 X 7	7,000	
Oil 50 litres X 10.00	500	
Miller's allowance 1500 X 12	18,000	
Depreciation	30,000	
Repairs	7,500	63,000

   PROFIT   37,000

(a)   If we wish to know how many kg per day should be milled to obtain this profit we have only to divide the number of kg milled in the year by the number of days it operates per year, thus,

$$\frac{100,000 \text{ kg}}{312 \text{ (operating days per year)}} = 320 \text{ kg per day (approx.)}$$

or again

320 X 26 (operating days per month) = 8,320 per month

But to cover the total expenses amounting to 63,000 UC

$$\frac{63,000}{1} \begin{array}{l}\text{(total expense)}\\ \text{(UC per kg)}\end{array}$$

63,000 kg should be milled per year

or 5,250 per month

or 202 kg per day

(b)   If the mill is depreciated over a period of 4 years, with annual depreciation of 37,500 UC, the miller's allowance fixed at 18,000 and the price of the milled rice at 0.75 UC per kilo, then annual expenses will be:

70,500 UC

and the annual income will be:

100,000 kg X 0.75 = 75,000 UC

with a profit of 4,500 UC

Therefore to cover these expenses $\frac{70,500}{0.75}$ UC per kg milled

94,000 kg should be milled per year.

ANNEXES PART VI

PROFIT AND LOSS ACCOUNTS

EXAMPLE OF AN ESTIMATED
PROFIT AND LOSS ACCOUNT
OF AN IRRIGATION PROJECT
for providing water to 10 hectares of riceland

FOR THE 4 MONTHS ENDING 31.12.87

Expenses	UC	Income	UC
Labour - Land preparation	30,000	Sale of 27,000 kg rice at 7.50 kg	202,500
Seed paddy	6,500		
Fertilizer (1st kind)	6,000		
Fertilizer (2nd kind)	6,000		
Insecticide / Pesticide	12,500		
Labour - Harvesting	12,000		
Threshing	7,500		
Total expenses	80,500		
Gross profit	122,000		
Total expenses	202,500	Total income	202,500

Practical exercice
ACCOUNTING OF A LOCAL ASSOCIATION

by Elhadji NDONG
Jacques MOYNAT

This exercice has been elaborated by 35 trainees in charge of the local management of their association. It is the result of their own experience in dealing with daily problems.

I. WORKING METHOD

1. Working groups by geographical areas

2. The leaders of these groups present the result of their work to all the participants

3. The "ideal" solution (discussed and adopted in a plenary session) is written distinctly on the blackboard and on the demonstration chart.

P.S. : The leadership of this seminar was in the hands of two qualified accountants, an expert and a leader of a local association.

II. OBJECTIVES OF THE EXERCICE

1. To help the participants to keep a cash or bank book correctly

2. To help the participants to register all the operations in an accounts and a bank book

3. To help the participants to enter the activities of each operation in the right account of expenses

4. To help the participants to follow up the financial evolution of their activities through a profit and loss account per activity

5. To help the participants to be self-sufficient: control of their different accounts, balance of movements, intermediary controls, etc.

INFORMATION DATA FOR THE EXERCICE

Activities of the Siripala Local Association
From July 1, 1988 to June 30, 1988

N.B. : The figures which are utilised do not always represent the reality. The purpose of this exercice is to understand the mechanism of allocation of the funds.

Situation at July 1, 1987

- Balance in cash :   125 UC

- Grindlay bank, Colombo : Current account :   23'800 UC
  Placement for renewal of the equipment :   18'500 UC

- Vehicle : an old Toyota

- Motopompe : a new one received in June 1987,
  financed by CIDA :   36'000 UC

Activities during the year:

4/7/87   Siripala goes to town to :

- 1 -   a) Withdraw 5'000 UC cash from the bank (cheque No 1001)
          b) Order 5'000 kg of fertilizer for rice production

          On return to the village he gives back to the cashier 4'980
          UC; the balance of 20 UC being used to pay the food in town

5/7/87   The cashier pays the village carpenter for making tables and
- 2 -   benches to be used for meetings : 475 UC

7/7/87   Fertilizer for the rice is furnished by a private company ;
- 3 -   the cashier gives them a cheque (No 1002) on the Grindlay
          Bank for 6'500 UC

8/7/87   Sirapala goes to town to buy different articles. On his
- 4 -   return to the village on 10 July, he gives the following
          statement to the cashier :

          - motopompe equipment :   660 UC
          - garden (wire-netting) :   3'570 UC
          - small computer :   180 UC
          - transport of equipment :   250 UC
          - food :   20 UC

He left the village with a blank cheque (No 1003) on the Grindlay Bank which he cashed in town in the amount of 4'800 UC. He has no more money on him.

12/7/87
- 5 -
The President gives to the cashier a letter from the Grindlay Bank informing them that half of the CIDA grant to construct the well has been allocated : 65'000 UC. He also gives him the Grindlay Bank's current account statement up to June 30, 1987. The balance of this account is 23'950 UC.

13/7/87
- 6 -
Siripala goes to town to buy the seeds for the production in the garden:

- seeds potatoes :	1'250 UC
- seeds tomatoes :	120 UC
- seeds cauliflower :	85 UC
- seeds onions :	240 UC

He left with 2'000 UC. Upon return he was unable to give any money back to the cashier.

14/7/87
- 7 -
Travel to Colombo to buy equipment for the wells. The expenses were the following :

- cement :	13'400 UC
- iron :	4'000 UC
- transport material :	800 UC
- food and bus :	250 UC

When he left the village for Colombo the cashier gave him an amount of 350 UC and on his return the treasurer made out a cheque (No 1004) for 18'000 UC extra to pay for what he bought.

15/7/87
- 8 -
The local association sells its Toyota car to a local trader who paid the Association by giving 10'000 kg of cement for the price of 4'500 UC. The cement will be used for the construction of the wells financed by CIDA.

16/7/87
- 9 -
Siripala goes to town to buy gas-oil for the motopompe. He buys two casks at 350 UC each and pays by cheque N°1005. He also paid 5 UC cash for transport which was repaid by the cashier.

17/7/87
- 10 -
The local association buys a new vehicle Datsun for 18'500 UC by cheque on the Grindlay Bank account "Placement for renewal of material and equipment". The insurance for the car will cost 1'600 UC which is paid in cash.

17/7/87
- 11 -
The members of the association begin to work in the collective garden. The cashier paid for food and drinks for the workers : 145 UC

18/7/87
- 12 -
The cashier repays to Siripala the supplementary amount that he spent when he was in town on 14 July.

20/7/87 - 13 -	Seeds bought for gardening. Paid by cheque No 1006 for 408 U.C. Tomatoes sold to the market : 6'500 paid cash by the clients.
1/8/87 - 14 -	Siripala buys 2 oxen to work with the CIDA well. The two oxen cost 1'200 UC, paid in cash. On return to the village, Siripala presents the following statement to the cashier :

- transport to town :   8 UC
- food :   20 UC
- transport back to the village : 16 UC.

When he left the village the cashier gave him 1'500 UC. On his return he gives the difference back to the cashier.

3/8/87 - 15 -	The Association buys a motocycle for its village animator. Paid cash : 1'600 UC
4/8/87 - 16 -	The President of the association signs a contract with OXFAM to finance a cereal bank (20'000 kg of millet) for the amount of 22'000 UC.
10/8/87 - 17 -	Siripala goes to Colombo and cashes cheque No 1007 for 17'400 UC to buy the following items :

- cement for wells :   4'500 UC
- iron :   2'300 UC
- transport of materials :   400 UC
- transport by car to town :   120 UC
- food and lodging :   40 UC

Siripala returns the balance in cash to the cashier.

11/8/87 - 18 -	Siripala goes 50 miles from the village to buy 5'000 kg of millet for the stock of the cereal bank. He pays cash, 5'100 UC, for the 5'000 kg
12/8/87 - 19 -	The Association decides to build a health centre. Siripala goes to town to buy 1'000 kg of cement at 470 UC. The wood for the carpentry is 68 UC. The straw 50 UC. The health centre is constructed by the youth of the association in two days. Their food costs to the Association 20 UC a day. Siripala had withdraw from the bank 10'000 UC. He gives the difference to the cashier.
13/8/87 - 20 -	Siripala buys medicines for the health centre: 857 UC. This amount is paid in cash after a withdrawal of 1'050 from the bank (cheque No 1009 current account). He paid for his transport : 63 UC and for food : 18 UC.

On his return to the village he gives to the cashier only 65 UC and signs a receipt for the difference that he will have to repay later.

17/8/87 - 21 -	The cashier buys accounting books for 25 UC.

18/8/87   The cashier receives in cash from Siripala the amount that
– 22 –   he had to repay to the association

31/8/87   31'000 kg of millet was bought for the cereal bank from
– 23 –   Wilpotha Inc. The invoice, including transport, will be sent
            in September : 13'200 UC. Withdrawn from the cash in Grind-
            lays Bank current account : 9'000 UC.

13/1/88   The Bank notifies that a credit of 7'500 UC has been
– 24 –   registered. It corresponds to the selling of rice.

14/1/88   – Withdrawn from bank in cash : 6'000 UC.
– 25 –   – Paid by cheque 1'450 UC for the invoice for car repair
            – The Association buys some equipment for the CIDA well :
               16'500 UC (paid by bank cheque No 1010)
            – Paid to Wilpotha Inc. : 13'200 UC, cheque No 1011

10/1/88   – The Association sells potatoes for 475 UC (cash)
– 26 –   – Pays salaries of personnel : 4'250 UC

30/6/88   The Association transfers to the account "Renewal of equip-
– 27 –   ment" at the Grindlay Bank the amount corresponding to
            10'000 kg of rice which was foreseen for the amortization of
            the motopompe : 6'500 UC.

                        * * * * *

*Please try to register all the operations
of this local association and compare your
work with the following pages.*

ACCOUNTING OF THE LOCAL ASSOCIATION

What to do and how to do it ?

1. Draw up the <u>financial balance sheet</u> (it is the same as the balance sheet at the end of the previous year)

   It should reflect the financial situation of the local association. It is not necessary to take into account investments and the list of items given in the inventory.

2. Establish the <u>Inventory of the equipment</u>. This is not an accounting document but a complete list of all equipment which is absolutely necessary to the group. This inventory is established from an extra accounting system at the end of the exercice.

3. <u>Create a cash book and a bank book</u>

   These books should register daily all operations of cash, movement of current account in the bank and placement of funds in the bank for renewal of equipment.

4. <u>Breakdown of financial accounts</u> into the separate accounts according to subject.

5. <u>At the end of each fiscal year</u> :

   a) Establish for each account the total of income and expenditure and determine the profit or loss for the fiscal year

   b) Establish a financial statement at the end of the exercice

   c) Establish the inventory of equipment which is the property of the local association at the end of the fiscal year.

NOTE :

For a clear presentation of the bank and cash book, it is important to consider :

- the date of each operation
- the number of the operation
- the item
- the amount
- the balance after each operation

The advantage of this system is to be able :

- to follow up the financial evolution of your activities
- to analyze and consider if your association is profitable
- to compare the different activities
- to compare your activities with other associations

1.   ENTRY FINANCIAL BALANCE SHEET UP TO JULY 1, 1987

ASSETS   LIABILITIES

Grindlay Bank : Current account	23'800	Financial capital	42'425

Grindlay Bank :
  Placement for renewal
  of equipment
	18'500		

Cash	125		
	42'425		42'425
	======		======

2.   INVENTORY OF EQUIPMENT AND INVESTMENT

UP TO JULY 30, 1987

	Item	Date	Value when new	Donor
1	Toyota	1983	13'500	Association
1	Motopompe	1987	36'000	CIDA

3. The different operations of this exercice are registered in the cash
   book as follows :

CASH BOOK

No doc.	Date	Item	IN	OUT	BALANCE
	1/7/87	Balance	125		125
1	4/7/87	Cheque No 1001 Grindlay Bank current account	5'000		5'125
1	4/7/87	Food		20	5'105
2	5/7/87	Table and benches for meeting	–	475	463
4	8/7/87	Received cheque No 1003 Grindlay bank	4800	–	9'430
4	"	Equipment for motopompe	–	660	8'770
4	"	Purchase of wire-netting	–	3'570	5'220
4	"	Purchase of small computer	–	180	5'020
4	"	Transport of material	–	250	4'770
4	"	Food	–	20	4'750
4	"	Siripala owes to the association	–	120	4'630
6	13/7/87	Potatoes seeds	–	1'250	3'380
6	"	Tomato seeds	–	120	3'260
6	"	Cauliflower seeds	–	85	3'175
6	"	Onion seeds	–	240	2'935
6	"	Siripala owes to the association	–	305	2'630
7	"	Received by cheque No 1004 from bank	18'000	–	20'630
7	"	Cement for the CIDA well	–	13'400	7'230
7	"	Iron CIDA well	–	4'000	3'230
7	"	Transport of material	–	800	2'430
7	"	Transport and food	–	250	2'180
7	"	The association owes to Siripala	100	–	2'280
8	15/7/87	Sale of Toyota	4'500	–	6'780
8	"	Purchase of cement for CIDA well	–	4'500	2'280
9	"	Transport of gasoil for motopompe	–	5	2'275
10	17/7/87	Payment of car insurance	–	1'600	675
11	"	Food and drinks gardening	–	145	530
12	18/7/87	Repayment to Siripala Colombo trip	–	100	430
13	27/7/87	Sale of tomatoes	6'500	–	6'930
14	1/8/87	Purchase of two oxen	–	1'200	5'730
14	"	Tranport and food	–	44	5'686
15	3/8/87	Purchase of motocycle for animators	–	1'600	4'086
17	10/8/87	Cheque Grindlay for cash No 1007	17'400	–	21'486
17	"	Purchase of cement for CIDA well	–	4'500	16'986
17	"	Purchase of iron for CIDA well	–	2'300	14'686
17	"	Transport of material for CIDA well	–	400	14'286
17	"	Transport, food and lodging	–	160	14'126
18	11/8/87	Purchase of 5000 kg millet for cereal bank	–	5'100	9'026
19	12/8/87	Cash cheque Grindlay Bank	1'000	–	10'026
19	"	Construction health centre	–	628	9'398
20	13/8/87	Grindlay Bank No 1009 for cash	1'050	–	10'448
20	"	Payment of medicines for health centre	–	857	9'592
20	"	Transport and food	–	81	9'511
20	"	Siripala should repay to the association	–	47	9'464
21	17/8/88	Purchase of office equipment	–	25	9'439
22	31/8/87	Siripala reimburses to the association the amount which he owes	472	–	9'911
23	"	Transfer from cash to bank	–	9'000	911
25	14/1/88	Withdrawn from bank in cash cheque No 1010	6'000	–	6'911
26	10/1/88	Salaries of personnel	–	4'250	2'661
26	10/1/88	Receipt cash for selling potatoes	475	–	3'136

3. The different operations of this exercice are registered in a Bank
   Book as follows (please refer to the items and to the dates of the
   operations)

BANK BOOK
Grindlay Bank, Current Account

No doc.	Date	Item	IN	OUT	BALANCE
	1/7/87	Balance	23'800	–	23'800
1	4/7/87	Withdrawn from bank to cash cheque No 1001	–	5'000	18'800
3	4/7/87	To buy fertilizer cheque No 1002	–	6'500	12'300
4	8/7/87	Withdrawn from bank to cash cheque No 1003	–	4'800	7'500
5	12/7/87	CIDA grant and contract	65'000	–	72'500
5	12/7/87	Interest of the current account	150	–	72'650
7	14/7/87	Withdrawn from bank to cash cheque No 1004	–	18'000	54'650
9	16/7/87	Cheque No 1005 to pay 2 casks gazoil	–	700	53'950
13	20/7/87	Cheque No 1006 to buy seeds for gardening	–	408	53'542
17	10/8/87	Cheque No 1007 withdrawn from bank to cash	–	17'400	36'142
19	12/8/87	Cheque No 1008 withdrawn from bank to cash	–	1'000	35'142
20	13/8/87	Withdrawn from bank to cash cheque No 1009	–	1'050	34'092
23	31/8/87	Received from cash	9'000	–	43'092
24	13/1/88	Transfer to bank. Selling of rice	7'500	–	50'592
25	14/1/88	Cheque withdrawn from bank to cash	–	6'000	44'592
25	14/1/88	Cheque car maintenance	–	1'450	43'142
25	15/1/88	Payment well equipment CIDA cheque No 1010	–	16'500	26'642
25	16/1/88	Withdrawn cheque No 1011 to pay 2 t of millet for the cereal bank	–	13'200	13'442
27	30/6/88	Transfer to the Grindlay account Placement for renewal of material	–	6'500	6'942

3.
BANK BOOK
Grindlay Placement for Renewal of Material

No doc.	Date	Item	IN	OUT	BALANCE
		Balance	18'500	–	18'500
10	17/7/87	Purchase of Datsun	–	18'500	
27	30/6/88	Transfer from Grindlay current account for the amortization of the motopompe	6'500	–	6'500

4. Breakdown : Donor account (well)

CIDA WELL

Expenditure   Income

Date	No doc.	Item	Amount	Date	No doc.	Item	Amount
14/7/87	7	Cement	13'400	12/7/87	4	Subvention CIDA	65'000
14/7/87	7	Iron	4'000				
14/7/87	7	Transport iron/cement well	800				
14/7/87	7	Travel and food costs	250				
15/7/87	8	Cement	4'500				
1/8/87	14	Purchase of 2 oxen	1'200				
1/8/87	14	Transport and food	44				
10/8/87	17	Cement well	4'500				
10/8/87	17	Iron	2'300				
10/8/87	17	Transport	400				
10/8/87	17	Transport and lodging	160				
14/1/88	25	Cheque No 1010 equipment	16'500				
		Yearly expenditure	48'054				
		Amount to justify	16'946				
		TOTAL	65'000			TOTAL	65'000

4. Breakdown : Donor account (cereal bank)

CEREAL BANK

Expenditure   Income

Date	No doc.	Item	Amount	Date	No doc.	Item	Amount
11/8/87	18	5 t. millet	5'100				
14/1/8	25	2 t. millet, cheque No 1011	13'200			Yearly loss	18'300
		TOTAL	18'300			TOTAL	18'300

4. Breakdown : Rice production activity account

RICE PRODUCTION

Expenditure   Income

Date	No doc.	Item	Amount	Date	No doc.	Item	Amount
7/7/87	9	Cheque No 1002 purchase of fertilizer	6'500	13/1/88	24	Sold rice	7'500
16/7/87	9	2 casks gazoil cheque 1005	700				
16/7/87	9	Transport gasoil	5				
		Annual profit	295				
		TOTAL	7'500 =====			TOTAL	7'500 =====

4. Breakdown : Gardening activity account

GARDENING

Expenditure   Income

Date	No doc.	Item	Amount	Date	No doc.	Item	Amount
10/7/87	4	Purchase of equipment motopompe	660	27/7/87	13	Sale of tomatoes	6'500
10/7/87	4	Wire-netting	3'570	30/1/88	26	Sale of potatoes	475
13/7/87	6	Seeds potatoes	1'250				
13/7/87	6	Seeds tomatoes	120				
13/7/87	6	Seeds cauliflower	85				
13/7/87	6	Seeds onions	240				
17/7/87	11	Food and drinks	145				
20/7/87	13	Seeds, cheque No 1006	408				
		Yearly profit	497				
		TOTAL	6'975 ======			TOTAL	6'975 ======

4. Breakdown : Interest account

BANK INTERESTS

Expenditure   Income

Date	No doc.	Item	Amount	Date	No doc.	Item	Amount
				12/7/87	5	Bank interest on statement of 30/6/87	150
		Yearly profit	150				——
		TOTAL	150 ===			TOTAL	150 ===

4. Breakdown : Health centre account

HEALTH CENTRE

Expenditure   Income

Date	No doc.	Item	Amount	Date	No doc.	Item	Amount
12/7/87	19	Building health centre	628				
13/8/87	20	Medicines health centre	857				
13/8/87	20	Transport medicines and food	81			Annual loss	1'566
		TOTAL	1'566 =====			TOTAL	1'566 =====

4. Breakdown : Expense account for the coordination of the association

GENERAL EXPENSES

Expenditure							Income
Date	No doc.	Item	Amount	Date	No doc.	Item	Amount
4/7/87	1	Food	20				
8/7/87	4	Food purchase of material	20				
8/7/87	4	Transport material	250				
8/7/87	4	Purchase of small computer	180				
17/7/87	21	Office equipment	25				
10/1/88	26	Salaries animators	4'250				
						Annual loss	4'745
		TOTAL	4'745			TOTAL	4'745
			=====				=====

4. Breakdown : Vehicles account

VEHICLES

Expenditure							Income
Date	No doc.	Item	Amount	Date	No doc.	Item	Amount
17/7/87	10	Car insurance (Datsun)	1'600				
14/1/88	25	Car repair	1'450				
						Annual loss	3'050
		TOTAL	3'050			TOTAL	3'050
			=====				=====

4. Breakdown : Investment / Equipment account

INVESTMENT / EQUIPMENT

Expenditure							Income
Date	No doc.	Item	Amount	Date	No doc.	Item	Amount
5/7/87	2	Tables and benches	475	15/7/87	8	Sale of Toyota	4'500
17/7/87	10	Purchase of Datsun	18'500				
3/7/87	15	Motocycle for animator	1'600				
						Annual loss	16'075
		TOTAL	20'575				20'575
			======				======

4. Breakdown :Loans account

LOANS

Expenditure							Income
Date	No doc.	Item	Amount	Date	No doc.	Item	Amount
10/7/87	4	Siripala owes to the association	120	14/7/87	7	Cash to Siripala	100
13/7/87	6	Siripala owes " "	305	18/8/87	23	Reimbursment to Siripala	472
18/7/87	12	Siripala owes " "	100				
13/8/87	20	Siripala owes from cheque 1009	47				
		TOTAL	572				572
			===				=====

4. Movement of funds account

MOVEMENT OF FUNDS

Expenditure				Income			
Date	No doc.	Item	Amount	Date	No doc.	Item	Amount
4/7/87	1	Withdrawal cheque 1001	5'000	4/7/87	1	Received cheque 1001	5'000
7/7/87	4	Withdrawal cheque 1003	4'800	7/7/87	4	Received cheque 1003	4'800
14/7/87	7	Withdrawal cheque 1004	18'000	14/7/87	7	Received cheque 1004	18'000
10/8/87	17	Withdrawal cheque 1007	17'400	10/8/87	17	Received cheque 1007	17'400
12/8/87	19	Withdrawal cheque 1008	1'000	12/8/87	19	Received cheque 1008	1'000
13/8/87	20	Withdrawal cheque 1009	1'050	13/8/87	20	Received cheque 1009	1'050
31/8/87	23		9'000	31/8/87	23	Received from cash	9'000
14/1/88	25	Withdrawal cheque 1010	6'000	14/1/88	25	Received cheque 1010	6'000
30/6/88	27		6'500	30/6/88	27	Received from Grindlay account "renewal for.."	6'500
		TOTAL	68'750			TOTAL	68'750
			======				======

5.a) At the end of the fiscal year : 30 June 1988
     Recapitulation of results (activities, etc.)

FINANCIAL RESULT OF THE EXERCICE

Activities	Total income	Total expenditure	Positive activities	Negative activities and running costs
Rice culture	7'500	7'205	295	
Bank interest	150		150	
Gardening	6'975	6'478	497	
Health centre		1'566		1'566
Running costs		4'745		4'745
Vehicle	____	3'050	___	3'050
Sub-total	14'625	23'044	942	9'361
CIDA well	48'054	48'054		
Investment/equipment	4'500	20'575	___	16'075
Sub-total	67'179	91'673		
Total loss of the year	24'494	____	24'494	____
TOTAL	91'673	91'673	25'436	25'436
	======	=====	======	======

5.b) Financial balance sheet at June 30, 1988

Expenditure   Income

Item	Amount	Item	Amount
Bank amortization	6'500	CIDA	16'946
Bank current account	6'942		
Cash	3'136	Capital 1/7/87	
OXFAM	18'300	42'425	17'932
TOTAL	34'878	TOTAL	4'878
	======		======

5.c) Inventory of equipment
up to the 30 June 1988

Item	Date	Value when new	Donor
1 Motopompe	1987	36'000	CIDA
1 Health centre	1987	628	Association
Tables - benches	1987	475	"
1 Motocycle	1987	1'600	"
1 Computer	1987	180	"
1 Datsun	1987	18'500	"

* * * * *

N.B.: *If this exercice corresponds to your own problems of management,
do not hesitate to utilize the same accounting system.*

* * * * *

ANNEXES PART VII

FINANCIAL CONTROLS AND JUSTIFICATION OF EXPENSES RELATED TO GRANTS

SPECIMEN OF ACCOUNTS DOCUMENTS SUBMITTED TO DONORS

Covering letter

WILPOTHA DEVELOPMENT ASSOCIATION

89th Milepost
Wilpotha
Sri Lanka

15th January 1988

CIDA Representative
P.O. Box 1234
Colombo

GRANT OF 160,000 UC FOR THE WILPOTHA WOMEN'S GROUP
SANITATION PROJECT 1987/88

Dear Sir,

We are pleased to forward herewith the audited accounts of the expenses of the 160,000 UC allocated to us for 1986 and 1987.

The following documents are also enclosed:

- a report of the activities carried out with this grant
- a summary of the financial assistance received
- a detailed account of the project expenses
- some annexes:

  the Association's accounts for 1986/1987
  the Annual General Report
  other documents, photographs, publications, etc.

We hope that this information will suffice.

We are at your disposal to answer any queries you may have.

We also take this opportunity to thank you for your support. We hope that you will work with us even more closely in the future.

Yours faithfully,

Siripala Perera
Chairman

Encl: A complete dossier

SPECIMEN OF ACCOUNTS DOCUMENTS SUBMITTED TO DONORS

Detailed financial statement

CIDA GRANT OF 160,000 UC FOR THE WILPOTHA WOMEN'S GROUP
SANITATION PROJECT 1986/87

UC

DESCRIPTION	BUDGET	ACTUAL EXPENDITURE		TOTAL
		1986	1987	
A. PURCHASE OF BUILD-ING MATERIAL	120,000			
- Cement		28,500	30,600	
- Sand		11,000	10.600	
- Bricks		13,000	14,300	
- Rubble for foundations		5,000	4,900	
- Ceramic squatting pan & tap (brass)		10,000	10,000	
		67,500	70,400	137,900
B. TRAINING OF BRICK LAYERS AND MASONS	20,000			
- Training costs and wages		8,500	9,800	18,300
CIDA TOTAL	160,000	76,000	80,200	156,200

Local Participation

Purchase of timber doors made with a part of the subscription fees amounting to 20,000 UC.

In 1986, trained masons were employed at a daily wage of 80 UC. Their total wages for one and a half months amounting to 7,200 UC were paid out of funds collected by a "Fund Raising Sale" by the Women's group.

Thirty young men received training in bricklaying and masonry work.

The original documents can be inspected at Wilpotha.

Siripala Perera
Chairman

SPECIMEN OF ACCOUNTS DOCUMENTS SUBMITTED TO DONORS

A summary of the financial assistance received

A. **PARTNERS**

   Name/Address of group: Wilpotha Development Association
   89th Milepost
   Wilpotha, Sri Lanka

   Donor: CIDA (Canada)

B. **PROJECT**

   Title of project: Wilpotha Women's Group Sanitation Project

   Amount granted: 160,000 UC

   Date of contract: 15th January 1986

C. **PAYMENT RECEIVED** Sampath Bank, Chilaw Branch
   Account No. 6172986-002

   Initial disbursement: 75,000 UC by transfer 20th January 1988

   Second disbursement: 60,000 UC by transfer 28th June 1988

   Final disbursement: 25,000 UC to be disbursed

D. **DOCUMENTS** refer to detailed accounts annexed

Summary of Project Expenses:	1986	1987	TOTAL
1. Purchase of building material	67,500	70,400	137,900
2. Training costs	8,500	9,800	18,300
	76,000	80,200	156,200

E. **ANNEXES** - Project Report

   - Financial Report

   - Newsletter of the Association in
   which this project is reported

Wilpotha 20th September 1988   Siripala Perera
                                                Chairman

SPECIMEN FOR AUDITING OF CURRENCY IN THE CASH BOX

CASH CONTROL FORM

DATE: *15/7/88*

In UC

No. of coins/notes	Value of coins/notes	Total amount
10 coins	5	*0.50*
10 coins	10	*1.00*
25 coins	25	*6.25*
10 coins	50	*5.00*
18 coins	1.00	*18.00*
10 coins	5.00	*50.00*
10 notes	10.00	*100.00*
12 notes	50.00	*600.00*
10 notes	100.00	*1,000.00*

Total available in cash box   *1,780.75*
=======

Accountable balance: *1,780.75*

Certified correct
(Signature of Accountant or cashier)

This exercice should be carried out each week.

SPECIMEN BANK RECONCILIATION FORM

This is the operation which reconciles balances appearing in the books and the statement of accounts sent by the Bank.

For example:

BANK RECONCILIATION

Sampatha Bank
Colombo

Account No. 30950618/B

Wilpotha Association

	Date	Amount UC
Statement given by the Bank	18.8.88	14,163.75
Statement according to our books	18.8.88	16,792.25
Difference		2,628.25 less in the bank

The difference is due to:

(entries not made)

15.7.88	CIDA balance monies received	6,700.00
16.7.88	Loan repayment to Bank	(3,500.00)
17.7.88	Petrol bill for June 88	(571.75)
		2,628.25

Accuracy of account certified - 18.7.88

Perera

K.D. Perera
Accountant

SPECIMEN FORM OF EXPENSE STATEMENT

Sarva Seva Farms Organisation
Kasayankulam
Tamil Nadu

EXPENSE STATEMENT

Name: R.S. Rajesh

Reason: Travel to Madras to purchase spare parts for
             the water pump.

Date: 11th January 1988

EXPENSES INCURRED		UC
1. Travel by bus (return trip)		16.00
2. Purchase of spare parts		570.00
3. Official procedures (stamp)		1.00
4. Miscellaneous: meals		25.00
	TOTAL	612.00

Cash received before leaving: 1,000.00

Total expenses   612.00

Amount which has been
returned today   388.00

Received the above amount :

Signature of member   Signature of Accountant

.....R.S. Rajesh....   ....Ple...........

Annexe enclosures and invoices

Kasayankulam, 13th January 1988

SPECIMEN INVENTORY OF GOODS
OF A DEVELOPMENT ASSOCIATION
as at 31st December 1988

	PURCHASE VALUE PER UNIT UC	PRESENT STATE	SUGGESTED ACTION TO BE TAKEN
A. MACHINES			
3 Mills	150,000	average	Replace Millstone
3 Engines	45,000	good	Maintain more often
B. TOOLS			
52 Watering cans	5,200	average	
20 Weeding machines	2,000	average	
20 Spades	100	need replacing	Buy new spades
25 Pick-axes	11,500	good	
C. PREMISES/FURNITURE			
5 Tables	etc.	etc.	etc.
25 Chairs			
3 Cupboards			
4 Boards			
D. VEHICLES			
1 Toyota Hiace van			
E. GOODS STOCK			
42 bags Cereal (Cereal Bank)			
10 bags rice			
20 bags peanuts			
	Purchase price or assessed price	good/average/ defective/ needs replacing	to be repaired to be replaced to be better maintained

A LIST OF THE NECESSARY CONTROLS TO BE EFFECTED WITHIN AN ASSOCIATION

CONTROLS

A. FINANCIAL

Control the Cash (accounts and available money).
Check if the bank balance is the same as stated in the accounts.
General check by an audit firm.

(Adapt rules if necessary)

B. ACCOUNTS

Justify every operation with corresponding receipt (voucher).
Check additions and balances.
Check Cash/Bank transfers.

C. STAFF/MEMBERS

Compare the work given to each worker/member, or group with the results obtained. Comment. Decide.
Check if the members have paid their annual membership fees.
Check their participation in community work.
Check if loan repayments are correctly made according to the contracts signed by the borrowers.

D. STOCKS (cereal from the cereal bank or spare parts, etc.)

Check if the quantity registered in the accounts is the same as that recorded after the physical inventory.
Check the available stocks and order well in advance.

E. VEHICLE/MACHINES/EQUIPMENT/TOOLS

Check their running condition.
Check their maintenance and decide on any action to be taken if necessary.
Check the validity of the estimated depreciation.
Calculate the cost per hour, per km.

F. WORK

Objectives of each project and results obtained.

FINAL EVALUATION

- Original objectives

- Present situation

- Decisions to be taken

- Future objectives

Note: Always ask the key questions: What? Who? Where? How much?
Why? For whom? To whom? For what?

EFFECTIVE AND CONTINUOUS CONTROL
LEADS TO EFFICIENT MANAGEMENT

HOW TO USE THIS MANUAL

1. Identify your problem with a KEYWORD

2. Consult the INDEX below

3. Refer to the PAGE or ANNEXE indicated

INDEX

GLOSSARY *

ACCOUNT: bank account, cash account where expenses or receipts are registered

ACCOUNTING: system of registration of all financial operations of a society or an association to determine the benefit or the loss

ADVANCE: an amount of money given to somebody who will repay later

ADDED VALUE: the value of the work involved to transform the item from a raw material to a finished product

AGREEMENT: the signed document between two partners (government and NGO, for example) to ratify decisions taken in common

ALLOWANCE: amount of money given to somebody for a certain work or to indemnify him

AMORTIZATION: accounting technique to allocate the global cost of an equipment to a number of years corresponding to its life

ASSETS: wealth of the association: equipment, machines, vehicles, buildings, stock, cash, placement in bank etc.

AUDIT: internal or external control of your accounting by an expert

BALANCE SHEET: statement of the situation of the wealth or debt of your association at a precise date

BILATERAL AID: aid given by a government to another government, from a country to another country

BILL: invoice for a small amount

BORROWER: person or group or enterprise which receives a loan from a bank or another person

BUDGET: estimate of expenses and receipts foreseen for a programme, a project, the association or an enterprise

CAPITAL: difference on the balance sheet between wealth and debt, between the assets and the liabilities

---

These definitions are purposely simplified in order to be easily understood

CASH BOX: a box where you put the money to pay small expenses

CASH-FLOW: money which is available in the bank and in cash, and which is ready for expenses

CASH POSITION: an amount of money that you have at a certain date

CHARGES (FIXED): the amount of expenses that you cannot decrease and that you should finance whatever your production

CLASSIFICATION OF ACCOUNT: numerical system to classify the account with a specific number

CO-FUNDING: system of co-financing between two agencies or partners

CONTRIBUTION (FINANCIAL): financial participation of the members of your association

CREDIT SAVINGS: amount of money that you are able to save and place in the bank

CREDITOR/SUPPLIER: the person, group or enterprise who gives you credit for items and who should be repaid later

CREDIT/DEBIT: double accounting system with credits on the right hand side and debits on the left hand side

CURRENCY (FOREIGN): U.S. dollars, Swiss Francs, Deutsch marks, £ sterling etc. are strong currencies and very often foreign currencies for Third World associations

DEBTOR/CLIENT: person, association or enterprise to whom you give credit and who should repay you later

DEBIT/CREDIT: double accounting system with credits on the right hand side and debits on the left hand side

DEPOSIT (FIXED): amount of money that you place in a bank

DEPRECIATION: amount of money that you decide to allocate for a year, for example, as an equivalent amount of the depreciation of the value of an equipment

DISTRIBUTION OF INCOME: how income is distributed among groups of people in a country or in an enterprise

DISBURSEMENT: amount of money transferred to somebody else from your bank or from your cash

DONATION: amount given as a grant to an association

DONOR AGENCY: governmental or non-governmental agency or foundation which is able to give impersonal aid to the Third World

EXCHANGE (RATE): the rate that you obtain in exchanging, for example, US $ for local currency

EXPENDITURE: amount of money certified by a bill or an invoice when you buy a service or a product

FINANCIAL RESOURCES: all resources with a financial character: cash, bank, cheque etc.

FINANCIAL REQUEST: dossier prepared for donor agency to request programme financing

FIXED ASSETS: capital equipment, building etc. that you cannot move

FIXED CHARGES: see Charges

FLEXIBLE FUNDING: global amount of money that you receive as a grant without being obliged to produce a project proposal

FOREIGN AID: aid coming from foreign countries

FOUNDATION: very often a donor agency which is able to give grants from the interest of its capital

FUND RAISING: techniques to be used by an association to raise funds to cover the expenses of its budget

FUND (RESERVES): amount of money that you are able to invest at middle and long-term for the survival of your association

GOODS (STOCK): the number of kilos of grain that you have in your cereal bank

GRACE PERIOD (CREDIT): period of a loan during which you have not to repay any capital or interest (after 3 years, for example)

GRANT: amount of money given as a gift to your association by a donor agency

GUARANTEE: being responsible for somebody else in relation to a bank or financial matters

HUMAN RESOURCES: the work of the population or you personally

INCOME: the result, in money, of your efforts and work production

INCOME GENERATING PROJECT: project which could generate some money for a group or a person

INTEREST: income from the capital or an amount paid into a bank

INVENTORY: list of all equipment, machinery, vehicles, stock etc. at a certain date

INVESTMENT: amount of money spent to buy equipment e.g. machines, vehicles or a building

INVOICE: the bill of an expense

JOURNAL: statement of expenses in cash, bank, etc.

JUSTIFICATION: technique to justify to a donor agency how you have spent the money

KEEPING ACCOUNTS: system of accounting and liabilities: the debt of the association and its capital

LOAN: amount of money that you request from a bank or from a friend and which you will have to repay

LOCAL FUNDING: money that you are able to raise locally in your village or your town

LONG TERM: more than six years

LOSS: when your expenses are more important than your income

MAINTENANCE: expenses related to the maintenance of an equipment or machine or another investment

MANUFACTURING PRICE: cost of production for making a table, for example

MARGIN PROFIT: amount of money that you foresee for your own benefit

MEMBERSHIP FEES: amount of money that each member of an association has to pay every year

MIDDLE TERM: two to six years

MULTILATERAL AID: aid given by the UN system to a country

NGO (NON GOVERNMENTAL ORGANISATION): a local association with non-profit making aims

NOTE (DEBIT-CREDIT): documents established by a bank to credit or debit your account

OPERATION: a transfer, a withdrawal from the bank in cash or any other accounting act

OPERATIONAL PLAN: planning of your project in a certain period of time

ORDER FORM (TRANSFER): letter sent to your bank to give an order to transfer an amount of money to somebody else

OVERHEAD COST: cost of all expenses for administration of your association

PETTY CASH: cash given to an employee to pay very small expenses

PERSONNEL MANAGEMENT: technique of recruiting, organising, administrating and evaluating the staff of an association

PLANNING: method of elaborating a plan for the development of your activities or your association or a country

PROFITABILITY: the yielding of advantageous financial results

PROFIT AND LOSS ACCOUNT: account where you register and centralize all the loss and profit of the different activities of your association to determine the general profit or loss for a certain period

PROFIT MARGIN: margin of profit

PROGRAMME: integration of a group of projects to be realised in a certain period

PROJECT: a specific activity that you intend to implement in the future

PROJECT REQUIREMENT: what it is necessary to foresee for the implementation of a project

RATE OF INTEREST: the rate you will have to pay to the bank when you ask for a loan

RATE OF EXCHANGE: see exchange

RAW MATERIAL: wood, grain, iron, etc.: product before transformation

REPAYMENT SCHEDULE: plan of repayment of your debtor

REQUEST (FINANCIAL): see financial request

RESERVES: amount of money put into savings for future use

REVOLVING FUND: fund constituted by the grant of a donor agency to an association and which gives this association the possibility to manage loans to its own members

RESOURCES: human or financial means of the association

RUNNING COST: expenses that you incur to run a mill, a vehicle etc.

SALE: price of a product you sell to somebody else

SAVINGS: amount of money that you have been able to place in a bank for your own security

SELECTION OF PROJECTS: operation made by a donor agency to decide if your project will be financed or not

SELLING PRICE: the amount of money that the client will pay to you or that you impose to your client

SHORT-TERM: less than two years

SOCIAL EXPENSES: expenses which are not related to the productive activities

STAFF: personnel of your association

STATEMENT (FINANCIAL): list of expenses to determine your situation at a specific date

STOCK: amount of kilos, litres or number of products that you have in your building

SUBSIDIZE: grant given by an administration or a government

SUMMARY OF EXPENSES: to group some expenses: to present a list of them

TERMS: see Short, Middle, and Long Term

TIME SCHEDULE: a personal planning of your activities

TRANSFER: see Order

VALUE (ADDED): see Added

VOUCHER: cash, bank or other document which establishes an expense

WITHDRAWAL: when you decide to go to the bank to cash a cheque from your account you withdraw this amount from your account

MANUAL OF PRACTICAL MANAGEMENT
for Third World Rural Development Associations

Volume I :
Organisation, Administration, Communication

CONTENTS

Part 3

ENVIRONMENTAL STUDY

A. List of problems
B. List of needs
C. Inventory of available resources
D. Identification of problems
E. Ideas for projects

ANNEXES

III.1 Technical Forms - Agricultural Management
      a) Inventory of Livestock
      b) Extent of cultivable land
      c) Family manpower
      d) Place where work will be carried out
III.2 The Reference Book of the Village
III.3 Organising Self-help, Self-reliance and
      Community Participation in Rural Development

Part 4

PLANNING, PROGRAMMING & PROJECTS

A. The planning process
B. Methods of planning
C. Designing a programme
D. Designing of projects

ANNEXES

IV.1   The activities of a programme
IV.2 (a) A practical example of planning
    (b) Creating jobs - A project planning guide for
        disabled persons in Africa
IV.3   Trial planning of a project
IV.4   Methods of programming a group of activities
IV.5   Plan of a programme designed for a rural community
IV.6   Programme to help villages of the UVA province
        towards self-management of community development

Part 5

THE ORGANISING AND FOLLOW-UP
OF ACTIVITIES

A.   Planning the use of available time
B.   Organising the work of the association
C.   Organising the running of the association
D.   Project follow-up
E.   The marketing of the produce

ANNEXES

V.1   *Documents which help the management of*
        livestock - cows
V.2   *Time schedule for cultivation of ten*
        hectares of rice
V.3   *Questions to be asked when carrying out*
        a market research

Part 6

EDUCATION, INFORMATION,
DOCUMENTATION

A.   Education and literacy activities
B.   Setting-up a small documentation centre or
     a rural library
C.   The newsletter of an association

ANNEXES

VI.1   <u>*Handbooks in Appropriate Technology and Development*</u>
         (a) Liklik Buk - A Rural Handbook Catalogue for Papua
             New Guinea
         (b) The Field Directors' Handbook - An Oxfam Manual for
             Development Workers
         (c) Example of a primary health care manual from ASSEFA,
             India
VI.2   <u>*Newsletters*</u>
         (a) IRED-Forum
         (b) NGO Management Newsletter
         (c) Example of newsletter of a local association
VI.3   *(a)* <u>*SATIS Classification Plan*</u> *for Appropriate Technology*
         (b) Appropriate Technology Centres - Members of SATIS

Part 7

COMMUNICATION TECHNIQUES

A. Addressing the public
B. Letter writing
C. Taking down notes
D. Organising and conducting meetings
E. Preparing reports and minutes
F. Audio-visual systems

ANNEXES

Part 8

THE ORGANISATION & ADMINISTRATION OF AN ASSOCIATION

A. Obtaining offical recognition
B. The members
C. The statutory functions of an association
D. The organisation of the secretariat
E. Personnel management

ANNEXES